Foragt for liv

To aktivister mod kolonialisme og apartheid kommet af dage under tragiske omstændigheder

Nekrolog

Af
PREBEN KAARSHOLM

TO NYLIGE begivenheder har i markant form understreget, hvordan kampen om magt og demokrati i det sydlige Afrika stadig præges af meningsløs brutalitet og foragt for liv – den ledende Zimbabwe-politiker Maurice Nyagumbos selvmord den 20. april og likvideringen af an-

Konsekvenserne af Willowgate-historien har været vidtrækkende. At to mindre betydelige ministre – Frederick Shava og Callistos Ndlovu – har måttet slippe taburetten og belave sig på en kommende retssag er måske ikke sensationelt. Mere bemærkelsesværdigt er det, at tre af de mest indflydelsesrige skikkelser i det post-koloniale hierarki også blevet tvunget til at trække sig tilbage.

Det drejer sig om forsvarsminister Enos Nkala, undervisningsminister Dzingayi Mutumbuka, og mest overraskende Maurice Nyagumbo, der var »senior minister« i regeringen og ansvarlig for integrationen af stat og partiapparat op til den forudsete indførelse af en ét-partistat i 1990.

Nkala og forsvinder og den

DEN MEST tragiske skæbne er Maurice Nyagumbos – indtil for få uger siden den tredje-mest magtfulde person i Zimbabwes regering. En gammel kæmpe og ikke nogen intellektuel, der havde været aktiv i nationalistbevægelsen, siden han som ung opgav en karriere som migrantarbejder i den sydafrikanske restaurations-branche og vendte hjem, som han levende har skildret i sin selvbiografi *With the People* fra 1980.

Nyagumbo havde ikke lutter venner i Zimbabwe – det var ofte ham, der i ZANU (PF) fik de beskidte jobs med at banke folk på plads, men han var respekteret som en mand af folket og for det, han selv havde udstået: Sammenlagt tilbragte han næsten 21 år i fængsel og interneringslejre, før Smith-regimet faldt i april 1980.

I slutningen af april nåede

Koevoet-specialisterne båsen og og likvidere og 300 forvirrede SWAP-rillaer i det nordlige Na som havde misforstå hele og troede, landet v at blive uafhængigt? mon O'Dwyer-Russell konservative britiske blad *Daily Telegraph* te en *Koevoet*-soldat i C boland, hvorfor så ma de dræbte fjender var på nært hold, i ansigtet synel adende efter at overgivet sig, fik han s

»De er SWAPO, og de de. Det er det eneste, tyder noget.«

Om morgenen d. sagde David Webster stitut for Socialantr ved Witwatersrand-u tetet i Johannesburg til sin kæreste Maggie man og tog ud for at lu hunde. Da han kom

The Humanities
between Art and Science

To Shula
from Preben
12 May 1989

Center for Research in the Humanities

Copenhagen University

The Humanities
between Art and Science
Intellectual Developments 1880-1914

Edited by
Michael Harbsmeier and Mogens Trolle Larsen

Akademisk Forlag
1989

The Humanities between Art and Science

Intellectual Developments 1880-1914

Edited by Michael Harbsmeier and Mogens Trolle Larsen

Copyright © 1989 by Center for Research in the Humanities

Akademisk Forlag
POB 54
1002 Copenhagen K
Denmark

Cover illustration: Alfred Kubin, *La grande tête*, 1899. Graphische Sammlung, Albertina, Vienna.
Cover layout: Uffe Rosenfeldt and Bente Sivertsen
Printed by AiO Tryk, Odense

ISBN 87-500-2783-2

Center for Research in the Humanities
Copenhagen University
Njalsgade 80
DK 2300 Copenhagen S
Denmark

Printed in Denmark 1989

Contents

Introduction

Mogens Trolle Larsen
Michael Harbsmeier

The Centre for Research in the Humanities in Copenhagen was set up in 1986 on the basis of a special grant to implement a broadly defined research program with the title "Cultural tradition and renewal"; it is supposed to further relations between the well established disciplines within the humanities in a narrow sense, but also to include certain areas which in Copenhagen traditionally belong in the Faculty of Social Sciences, first of all Anthropology and the Copenhagen specialty called "Cultural Sociology".

This openly inter-disciplinary approach, which obviously aims at a certain softening of the rigid boundaries which have been erected between the various disciplines, sub-disciplines and specialties, made it natural for the first members of the Centre to start a reflection on the question of the history and nature of precisely those boundaries: when were the various "disciplines" created as distinct units, what were the intellectual, social, cultural or maybe political factors which furthered the process, and how did the various disciplines develop their "identity" in relation to each other?

Questions of this nature formed the basis for the decision to convene an international congress which should attempt to describe and analyse the intellectual developments in the crucial phase in European history, the *fin-de-siècle* period between roughly 1880 and the First World War. We intended to look at a whole series of individual disciplines and to invite scholars at home in different national traditions in order to gain as broad a perspective as possible. This volume contains the papers which came out of that conference which was held at Elsinore on December 5-8, 1986.

The title of the conference - and this book - was inspired by the seminal work by Wolf Lepenies, "Die Drei Kulturen", published in 1985, which has the subtitle "Soziologie zwischen Literatur und Wissenschaft". In this book Lepenies deals with the three national traditions: France, England and Germany, but the three "cultures" also refer to the complex relationship between the two traditional areas of Literature and Science on the one hand, and the new discipline of Sociology on the other, an intellectual investment which had to find its methods, logic and social place in a field carved out between Art and Science. Precisely in the first years in the life of this new discipline the complexities in the situation were felt to be particularly acute, where the three different approaches competed as possible, meaningful systems of explanation and interpretation of the human condition. We intended to point to this problematic area which seemed to us to constitute a fruitful basis for a better understanding of the intellectual developments around the turn of the century in the many disciplines in the humanities.

It is clear and well recognized that this period marked the professionalisation of the humanities. Some disciplines were based directly on traditions which had already built up a large body of ordered knowledge and a basic demarcation in relation to other pursuits; however, new concerns regarding methodology as well as the proper definition of the object of study came to the fore, for instance in such a relatively well-defined field as history. Other disciplines, such as sociology or psychology or structural linguistics, were much more radical departures which only to a limited extent built on the existing traditions.

Those who are accustomed to regard their own scholarly work as growing out of a centuries long tradition, which has led to a rational and meaningful division of work among logically demarcated fields, should consider the examples provided here of precisely how the concerns of today are shaped by the fights and division of interests which marked the period under discussion. In a particularly clear-cut example Stoklund shows how the split between traditional history, history of mentalities and cultural history has its roots in debates which took place around 1900 among German historians. Essential differences between French, German and British schools of sociology may be traced back to the way in which the shared intellectual traditions of "the history of European culture and institutions and a sense of its profound importance to the understanding of all social life" were transformed in the context of political and social differences in the three countries - as shown by Burrow. Or, to take a last example, the tiny discipline of Assyriology underwent severe convulsions in connection with fights over the proper understanding of religion (Christian and Jewish), and these battles left scars in the form of unapproachable black holes, themes and questions which could no longer be discussed.

The process of academic professionalisation and disciplinary diversification is a well-known feature of the period, of course, but it seems that there are certain trends which are shared by many or all of the disciplines, whether new or old.

Freud´s somewhat uncomfortable claim that "sometimes a cigar is just a cigar" and not a symbol of some unconscious desire or repression, mirrors the more general discovery of structures which lie beneath the surface, which in fact shape "reality" (*langue* reveals itself in *parole*) and which can be studied. Once this is realised, a war or a picture or social life or a speech can no longer be thought of as "just" that - in fact all such phenomena are reflections in some sense of a set of structures which become the subject of disciplines such as sociology, psychology, art history, literary studies etc. It may even be meaningful to place in this context the concerns with the masses, with the structures of everyday life, with barbarism and the exotic - topics which exercised the imagination of our *fin-de-siècle* predecessors.

These efforts take on the claim to be "Scientific Projects"; Freud´s own "Wissenschaftliches Projekt", in which he attempted to marry the study of the soul to hard science by creating a theory of neural psychology, was of course never made public, but it formed the basis for his entire later production (Wollheim ..); the conflict alluded to in our title is characteristic of for instance Polish sociology as described by Szacky or the work of Dilthey which is touched upon in several chapters in this book. The development of Znaniecki from his position of denying any kind of sociology access to the most important social facts, those which determine "creative development", to his later life as a Chicago sociologist, is highly revealing; his way went *via* culture which became the "harbinger of a new paradigm of all knowledge" (see page 32), and which enabled his vision of sociology to occupy a legitimate place among the scientific disciplines.

Finally, the political realities of *fin-de-siècle* Europe, socialism, imperialism and nationalism, can of course be directly traced in the history of our disciplines - from history and sociology to assyriology and art history. The common ground was a Eurocentric view of the world, where the domination of the West remained fundamentally unquestioned, and where the European experience became the yardstick with which all human phenomena could be measured. The classical stance of three African women posing as the three Graces for a photographer is revealing of the blindness which this cultural baggage imposed on the western traveller (see Kramer´s chapter), a fact which is obviously at the

Mogens Trolle Larsen and Michael Harbsmeier

root of the Orientalist discourse as well as the concern with Europe's own past. At the same time, and building on such premisses, these same scholarly disciplines played a most vital and active part in dismantling them and inaugurating a much more acceptable discourse.

The many different starting points represented in the papers in this book - in terms of national traditions and disciplines - lead to a kaleidoscopic view of the period - although one which reveals clear trends, repetitions, connections, and shared intellectual roots. We have had no ambition to present a coherent theory which could unite us all - the time is hardly ripe for that - but the papers in this volume offer doors which allow us intriguing insights into the gardens where our disciplines were first cultivated.

Welcome Address: The Humanities between Art and Science

Ove Nathan, Rector magnificus, Copenhagen University

To invite a scientist to give a welcome address to a conference on the humanities may seem somewhat inappropriate. What does the blind know about the colours? But since the theme of the conference touches upon the relation of the humanities not only to art, but also to science, I shall permit myself to use or abuse the opportunity and present some scattered thoughts on the scientific situation around the turn of the century - with a view to the implications for the humanities, taken in a broad sense.

Indeed, a lot of things happened in science in the period from 1880 to 1914. Let us count Darwin as a part of this period, since anyway he lived until 1882. His *Origin of the Species by Means of Natural Selection* from 1850 - still in 1880 was considered by many people as an infamous work, a threat to Christianity and human dignity. Darwin undermined the idea of a static world into which plants, animals and humans were brought once and forever perhaps by a singular event of creation. The living world became dynamic, and the human being no longer - in terms of biology - could be given an exclusive position, in splendid isolation from the animal world! Darwin's hypothesis further raised a fundamental question, as to the microscopic mechanism hidden to our eyes, behind the game of evolution. An intonation to the subject in fact was made in the middle of the 19th century, but apparently the hint was only understood by the turn of the century. I think of Mendel's work, his simple crossing experiments on strains of flowers of various colours. Slowly we learned about the laws governing heredity in biology, and about the phenomenal stability inherent in what we to-day would call genetic material. But the road to a deeper understanding of biology, the road to microbiology had to await for another turn of events in the sciences, namely in physics.

Radioactivity was discovered by Becquerel in the eighties, but its statistical nature remained a mystery to the Newtonian and

Maxwellian, so-called classical physics. Here again was a strange phenomenon awaiting a microscopic description. More conflicts with Newtonian thinking came up in 1905 when Einstein published his theory of special relativity. Fundamental concepts such as time-intervals, length and the notion of simultaneous events suddenly lost their absolute sense. Although Einstein's theory did not at all invalidate classical concepts of determinism, I think that to many laymen outside of physics, special relativity became another symbol, together with evolution, of a revolution in the sciences. The apparent world was in a state of dissolution, you could no longer believe in what your eyes and ears directly would tell you.

And more was about to come. As a minor step, the English physicist Rutherford could resolve one of the puzzles in the Darwinistic hypothesis. Using measurements on radioactive rocks Rutherford could prove, I think it was in 1912, that the life span of the Earth and thus the Solar system had to be enormous. Hundreds of millions, perhaps billions of years, enough for Darwinistic evolution to take place on Earth. This was another torpedo against the simple, mechanical model of the world belonging to classical physics. Giving rise to still more questions: What can science say about the evolution of the universe, the origin of life, the special position of our planet in the world?

And finally, perhaps most shocking of all, in 1913, the young Niels Bohr's startling picture of the hydrogen atom, built on the so-called quantum hypotheses. Here, for the first time, a little bit of the physical micro-world was penetrated by man. And it was quickly understood that in this micro-world we have to give up classical determinism, the concept of the world as a classical machine. Determinism in a statistical sense replaced the old notion, and with this physics brought into science and philosophy a new set of epistemological problems.

Bohr's atom is stable in an extraordinary sense. This quantum mechanical stability of the fundamental building-block of matter has far reaching consequences. When molecular biology evolved 40 years after Bohr's discovery, strange giant molecules called DNA and RNA were found to contain the genetic message. The stability of these molecules guarantee the stability we know so well in the usual reproduction of living beings. This molecular stability of DNA and RNA has the same quantal origin as the stability of Bohr's hydrogen atom. But small changes may happen from generation to generation in the genetic coding. The rare phenomenon of mutation can occur, and the road to a full microscopic explanation of Darwin's evolutionary processes perhaps one day may be found.

My reasons for throwing these scattered remarks into the beginning of your discussions on the humanities are, at least, twofold. The first point I wish to bring up is the question: can science replace the humanities? The second: can science inspire the humanities? Concerning the first point let me first summarize my previous remarks by emphasizing that also in the sciences the time span from 1880 to 1914 was full of changes and discontinuities. Revolutionary ideas were given birth, resulting in a drastic change in our scientific *Weltbild*. Whether these technological changes and especially their associated technological developments have been and will be for the benefit of mankind, or the opposite, is a question which divides people. No simple answer can be given to that question, as I see it. This is partly because technology is a result not just of science, but of the complicated interaction of science and society. Science and technology can be used but also abused. Anyway, the many changes and revolts which happened in science at the turn of the century have had an enormous impact on the modern world, and I think that to many laymen this was a period where science so to speak took over and replaced religion.

Let me dwell a moment at this point. To many people modern science has the character of a majestic cathedral forming an entity within which explanations are available to almost any question. This feeling is supported by the rapid and almost triumphal developments in the sciences which have happened in our times since the turn of the century. True enough, science has eliminated some previous ignorance. But we are still far from having a coherent, scientific Weltbild which brings together into one single picture the smallest elementary particles, the atoms and molecules, the living cell, the many-cell organisms, not to speak of the processes of the brain in human beings. You cannot explain human psychology and not even the properties of the most primitive animal from the laws of quantum mechanics. In my view one should have a much more modest view of science. As a method of working it is of tremendous power. It has built-in technological perspectives which we still can only guess about. But as a coherent Weltbild it is still only a fragment, and maybe it will remain so forever. It has removed a few cubic millimetres from our sea of ignorance, but science still leaves us with infinitely more questions than answers. Science cannot decide the fundamental questions of religions for us, it cannot replace the humanities.

This brings me to my second reason for talking so much about the sciences at the turn of the century. Even if modern science can throw light only upon a limited part of our universe of ignorance, it may perhaps contain elements and fundamental approaches which could be interesting and inspiring to the humanists, perhaps sometimes even become helpful.

On this I do not feel competent to give a long and learned talk, but just a couple of illustrations.

I have mentioned several times the scientific problem of describing macro-phenomena in terms of micro-phenomena. Can Darwin be understood in terms of molecular biology, DNA and all that? Can the properties of a piece of metal be understood in terms of atomic structures and substructures? Can the development of the entire universe be understood from the properties of tiny elementary particles?

Similar problems exist within the humanities or on the borderline between the humanities and science. What are the relations between the collective behaviour of society and the behaviour of the single individuals? In other words what are the aspects of individual psychology which are specifically relevant for sociology? Or you may turn to linguistics and ask questions as to the possible relations between the grammatical description of language - the macro-level - and the neurological phenomena of the brain - the micro-level. Perhaps the linguist and the neurobiologist never will be able to bridge the abyss between them. But, as has been pointed out by Wolfgang Dressler in a recent article on Complementarity in Linguistic Observation, Description and Explanation, perhaps certain useful parallels anyway may be drawn between physics and linguistics. Perhaps even the linguistics may learn something from Bohr's notion of complementarity in this respect.[1]

In this connection one should emphasize that modern science certainly is strongly linked to the humanities through our use of a common language. Bohr once said that physics deals with what we can **say** about the world. Science must use a language, and you must be careful in using your language, or you arrive at wrong

---

[1] Wolfgang U. Dressler, Complementarity in Linguistic Observation, Description and Explanation. In: *The Lesson of Quantum Theory*, edited by J. de Boer, E. Dal and O. Ulfbeck, North Holland 1986, p. 315.

conclusions. The language so to speak represents a meeting place between the sciences and the humanities.

These remarks carry me naturally from the turn of the century right up to the present. My point is that the humanities in the recent years are being confronted with science or rather with science-based technology in a new way - where language plays a central role. The computer is no longer a tool just for physicists, economists and business men. Its services are now being offered to the humanities, too, and apparently not just in trivial ways. Text editing by the individual researcher has taken a radically new turn. Data bases from foreign research centres become internationally available even within such fields as classical philology, and the life of lonely humanistic research groups thereby may change radically. And more profoundly, computer science now develop into areas such as expert systems, artificial intelligence and machine translation. Here the computer comes directly into interplay with humanistic fields such as linguistics, phonetics, philology, psychology, and philosophy. The computer may somewhat change the content and the quality of these humanistic fields. Whether this is for the good or the bad, I dare not say. But I think it is safe to say that more of this is going to come, and the impact on the humanities will probably grow in the coming years. There are limits to the power and possibilities of the computer, but we may be far from having reached these limits yet. Personally, I am on the modest side. I don't believe in the school without teachers, or in the perspective of an intelligent computer that can walk into a lecture room, sit down and listen, and then start a spontaneous discussion on a subject such as the humanities between art and science. But I do believe that a lot can be learned also on the humanistic side by a symbiosis with the computer scientists. The danger is, of course, that the computer with its rigid requirement to language, may restrict rather than enrich a field. Some people, even humanists also get so fascinated by computers that they stop thinking by themselves. However, I believe this is a sort of infant disease in the beginning of the computer age, which will disappear when we all mature with the computer.

Let me try finally to justify my use of so much of your time on the periphery of your subject. It is true that the points of contact between the humanities and the science in many fields are sparse, and that one shouldn't over-emphasize the importance of such contacts. Intuitively, however, I feel that the matter is important and deserves our attention. The methods, paradigmas and approaches and the language of the humanities and the sciences often differ substantially. But the fundamental creative processes involved in the two cultures, if I may use this expression, may not be that different. In our efforts to discover previously hidden relationships we are driven by a common curiosity and some ill-defined desire for beauty. Perhaps a deeper analysis of such creative processes will show that the distance is not all that great between the humanist and the scientist, and perhaps not even between their worlds and the world of the artist. In that sense we should perhaps not visualize the humanities as being "in between the sciences and the art":

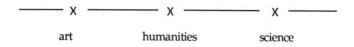

and approaches... art   humanities   science

but rather see all three activities as expressions of a common human desire for creativity, and therefore represent them as three points on a circle:

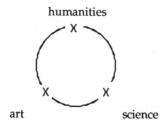

humanities

art   science

- all three individually being between the two others.

So let me end by the old question: are these three different cultures - or just one and the same?

Coherence and Specialization in the Social Sciences in the Early Twentieth Century

J.W. Burrow

I have taken, I am well aware, a large and ambitious subject, and I have to begin by freely admitting that what I shall offer will be a subjective and highly selective version of it, guided by perspectives of whose limitations I am no doubt not fully conscious. I take it that one of the functions of an interdisciplinary conference, focussed on a particular period of European culture, is to expose and exchange the mental images or topographies of the past, with all their individual, national and disciplinary limitations and biases, which we all possess. Hence this lecture is an attempt to offer mine, for the period with which we are concerned; not, certainly, as in any sense authoritative but as something more like a confession; this, at the outset of this conference, is where I personally have to start. Moreover, the limitations in my representation of the period, if unavoidable, are in a sense entirely appropriate to my general theme and epitomize what I shall attempt to speak about. Because my theme is comprehensiveness and specialization and the prices we pay, in terms of the former, for the advantages, if they be such, of the latter.

Perhaps it is in some degree the perennial human disposition to idealize the past which induces us to see past periods of scholarship as those of great polymaths, of men who could comprehend, if not the whole, then, compared with our own more minute specialization, a vast amount of the thought and learning of their times. If we think of the beginning of this century it is tempting to cast Max Weber, economist, jurist, historian, sociologist, thwarted politician, philosopher of social science, in this almost Faustian role, though the nineteenth century as a whole offers even more striking examples of the aspiration to comprehensive knowledge: Hegel, Marx, Comte, Herbert Spencer. But even when we think of lesser figures it is tempting to envy what often seems an easy, even unselfconscious transgression of the boundaries of disciplines now hedged with distinctive and forbidding technicality, and to envy too, perhaps, the common classical matrix, the grounding in the literatures of Greece and Rome, however superficial it may sometimes have been, which provided something like an elementary international culture for the educated classes of Europe. Of course, this nostalgia, if we do feel it, does rest in part on the illusion provided by distance, in which only the great mountain-peaks are clearly visible: and we easily ignore, because they have slipped from sight and memory, except, sometimes, within the confines of a particular discipline, the accumulations of fragmentary, specialized data which then as now consumed whole lifetimes of scholarship.

But surely, we may say, if we go back far enough - and perhaps we do not have to go as far as Aristotle or even the Renaissance - the illusion of possible comprehensiveness becomes, if not a common reality, at least a closer approximation to it. In the mid-eighteenth century, when D'Alembert, building on the foundations laid by Bacon and Locke, set out for his own age, in the *Discours préliminaire* to the *Encyclopédie*, the anatomy of all human knowledge, the enterprise of clothing it with flesh already seemed necessarily a collective one; yet Rousseau and Hume, Adam Smith and Goethe were to combine high originality and polymathic activity to a degree hard to conceive of not only now but perhaps also, already, by the beginning of this century. Hume was at once historian, epistemologist, psychologist, political scientist, philosopher of law and morals. Smith, lecturing at various times on Moral Philosophy, Jurisprudence and Rhetoric, covered the whole continent on which were to establish themselves the later distinct kingdoms and territories of the various sciences, as well as laying a large part of the foundations of one in particular, political economy. As I said earlier, we can find comparable examples of comprehensiveness of aspiration in the early and mid-nineteenth century. Smith and Marx have it in common that the works for which they are famous represented, for themselves, only a fragment of a much vaster, unfulfilled intellectual project, immense syntheses of which we have only the shadowy outlines.

Yet already, of course, we have in the eighteenth century lamentations, like that of Friedrich Schiller (Schiller 1958: 582-83), over the supposed fragmentation of the cultural world, as well as assertions, of which the prime example must be Herder's of the validity and perhaps, even, ultimately it sometimes came to seem, the incommunicability of distinct national cultures. It is precisely at this point too that we get the notion of individual comprehensiveness, not merely of knowledge but experience, formulated as an ideal, and the beginnings of the wistful idealization of the manysidedness, the universal interests, of Renaissance man, which received their definitive formulation in the picture presented by Jacob Burckhardt in the mid-nineteenth century.

If we turn finally, however, to the later nineteenth and early twentieth centuries, we find, I think, a new cultural variation on the theme of comprehensiveness; comprehensiveness sometimes seen not just as an ideal, an opportunity, or a lost human possibility; not exactly as Goethe's or von Humboldt's optimistic, neo-Hellenic idea of *Vielseitigkeit*, though it owed something historically to it; but rather as a cultural experience, defying attempts to wield it creatively, to make anything new or distinctive. It was a vision of European cultural life as a kind of museum-curatorship, exhausting its energies in mastering its own past: an Alexandrian culture, condemned, in the pursuits of history and criticism, and in the pastiche of artistic 'revivals', to express its creativity, at best, in imaginative re-creations of the past, and, at its most sterile, in increasingly minute and pedantic scholarship. Justified or not, that was certainly one self-image of the age; that sense of cultural exhaustion that contemporaries in large part meant when they spoke of their own as an age of decadence. It was a sense which sometimes prompted analogies, in writers as different as Georges Sorel and Walter Pater, with the later Roman empire, and gave a fashionable zest to vague notions of Wagnerian revival or transcendence, even, if necessary, at the cost of a kind of barbarism, a Gothic invasion *à l'intérieur*. The comfortable, outwardly high minded, middle-class museum culture, a massive exercise in possession, the culture of Nietzsche's *Bildungsphilister*, seemed to be at once the legacy of historicism, of an overdeveloped yet also ultimately belittling sense of the past, and the enemy of vitality, creative energy and will.

I shall use the term historicism a good deal here, and it is notoriously, in English, a term without a fixed meaning, while some of those it has are at odds with the sense of '*Historismus*', from which it derives. I shall use it here with a deliberate looseness, to indicate a general overall conception of human or at least of European history employed as a matrix of one's thinking, and sometimes, though not invariably, containing the Hegelian notion of an accumulation in which nothing is lost, but each stage of history incorporates whatever of value preceded it, proceeding (perhaps) to a final, comprehensive consummation. I use the term, as I say to include this notion, but also more loosely, in a sense compatible with the more positivistic versions of human history for which, in the later-nineteenth century, the phrase 'social evolution' became the fashionable term. It does seem, in fact, that the Hegelian Idealist historicism which pointed to a total cultural comprehensiveness as the culmination of history sometimes acquired in this period a suggestion of the doom-laden and the apocalyptic. The enervation that is the effect of overwhelming accumulation had been expressed in that famous image of a world-historical cultural comprehensiveness, Leonardo's *Giaconda* as seen by Walter Pater. Pater's essay, published in 1869 (Pater 1873), was in part an expression of the newly awakened interest in Hegel at Oxford at that time, but the period of Pater's greatest influence was the closing years of the century; mingled with the fashionable *fin de siècle* notion of the physical, biological degeneration or enervation of human breeding stocks - Darwinian evolution turning back on itself, as it were. The notion of a total possession of the culture of the past forms a part of that rival image and *reductio ad absurdum* of decadence, the hero or anti-hero of Joris-Karl Huysmans' *À Rebours*. It was the oppressiveness of the historical ideal of comprehensiveness, of the Hegelian Absolute-as-nightmare, which gave its exhilaration, its sense of cleansing, of shedding a vast accumulated weight, to the early experiments in literary and artistic modernism, to the insistence on allowing the literary artefact to make its own meaning, and which gives its paradoxical vitality to a celebrated modernist poem ostensibly about sterility - to the epitaph for European cultural coherence, an exploitation both of accumulation and fragmentation, which seems both to lament and to celebrate its emancipation: T.S. Eliot's *The Waste Land*.

At the risk of labouring the obvious I have dwelt on the theme of coherence, comprehensiveness and fragmentation in the European culture of the later-nineteenth and early-twentieth centuries, as a preamble to considering the social sciences in this period, because it seems to me that it constitutes an indispensable context for understanding the latter. The sense of cultural exhaustion or over-ripeness expressed in the idea of decadence, often associated with the idea of an all too well preserved and studied past, seems in marked contrast to the hopes that were still characteristically invested in the natural sciences, in the excitement of scientific discovery and the practical and theoretical novelties that the natural sciences seemed endlessly capable of producing. This contrast between what have since come to be christened 'the two cultures' widened, it seems to me, the rift which had been apparent at least since the era of Romanticism, and gives a peculiar retrospective interest to the history of the social sciences, poised, in fact if not in intention, 'between art and science'. The social sciences, or writing about social science, often exhibited, far more than ever before, in this period, the sense of occupying debatable ground. Some writers on society or the human sciences - I would be particularly inclined here to instance Sorel and William James - testify to a sense that the nineteenth-century historicist heritage, in philosophy and social and political reflection, is oppressive, just as restlessness with its apparently deterministic implications could lead to at least an equivocal attitude to the concept of science itself, in the work of Mach, Vaihinger, Poincaré, leading in the long run to an acuter sense of creative role of the the scientist in the formulation of hypotheses and conceptual schemes (Hughes 1958: 108-112). It would be harder in future to think or speak of science, as T.H. Huxley had done, as 'generalized common sense'. The world of the natural scientist would come to seem increasingly divorced from the simple empiricism which covered all accurate scholarship and observation with the benediction 'scientific' and hence seemed to make the creation of the human sciences little more than a matter of scrupulous caution and trained observation - the conception of social science embodied in England, for example, by the work of Sidney and Beatrice Webb.

Yet if the adjective 'scientific' was cherished, and still seemed for the moment available on fairly easy terms (any account of its popular usage during the period, for example, would certainly have to take account of the claims of Sherlock Holmes), 'science' in its more rigidly deterministic aspect could seem oppressive as well as inviting. There were psychological parallels in fact, between the closed world of an historicism for which all possibilities seemed both realized and exhausted, and the closed world of scientific determinism. The immense vogue during the period for the philosophy of Bergson, for example, is surely due largely to its skill in combining elements often seen as incompatible: an historicism (spoken of as 'evolution') reconciled with indeterminism and creativity: *l'évolution créatrice*. Evolutionism taken *à la Bergson*, as one finds it in England, for example, in Oscar

Wilde and Bernard Shaw (Wilde 1910: 384; Shaw 1901) allowed one, as it were, to have one's cake and eat it. On the one hand it spoke the language of the most fashionable and, for human beings the most apparently relevant of the natural sciences, evolutionary biology, and incorporated into it, by adopting a neo-Lamarckian version also currently fashionable, a notion of collective race-memory which was, in fact, another version of the ideal cultural comprehensiveness of the Hegelian historicist. Yet on the other hand it presented an indeterminist and open-ended version of it, incorporating the possibility of future creative, collective, self-transcendence, in contrast to the completed utopias which the classic versions of nineteenth-century historicism had typically presented as the climax of history.

So far I have spoken of coherence and of fragmentation rather than of specialization as such. It is in the social sciences during this period - the *parvenus*, as it were, of the academic community - that specialization and, what amounts to much the same thing, professionalization, is most obvious. Yet to some extent the same is true of the humanities generally. Although the academic establishment of some subjects - law, history, philosophy - reaches far back, this is above all the age of academic professionalization and its ramifications: chairs, academic journals, university courses and degrees, examinations, doctorates, academic clientage and discipleships, and vocal academic self-justification in terms of expertise, training, social and national usefulness, and the research ideal (Clark 1973; Rothblatt 1976: 164-73). With these things, as cause and consequence, goes a greater measure of specialization, of distinct disciplinary traditions, technicalities, perhaps even something resembling Kuhn's scientific paradigms. Emile Durkheim, in his first published review, in 1885, said that sociology had emerged from its heroic age (Lukes 1973: 85). He meant, presumably, that it was ceasing to be the product of isolated great thinkers, devising total sociological systems - Comte would, obviously, be the archetype - cut off from the stimulus, the criticisms and the empirical contributions of colleagues, that university departments exist, among other things, to provide. If that was so, then the sociologist no longer needed to be a genius or a kind of prophet, though Durkheim, as we know, took his own role as moral and political instructor very seriously; sociology could become the co-operative, empirical, continuously advancing intellectual enterprise for which the natural sciences provide the model.

We can see a similar kind of thought in the conception of social science that guided the founding parents of the London School of Economics, Sidney and Beatrice Webb, a decade later (Caine 1963). The School was to train its students in the craft of 'social investigation'. University departments and degrees are obviously one important aspect of what I am pointing to; another is the foundation of academic journals, with their implicit - and perhaps *only* implicit - assumption of the possibility of ultimate syntheses of the piecemeal results of individual research. The academic journal is surely as characteristic and significant a cultural product of this period as the great general reviews like the *Edinburgh Review* or the *Revue des Deux Mondes*, read by all educated people, were of the early and middle years of the last century. In England, the premier journal of the historical profession, the *English Historical Review*, was founded exactly a century ago, in 1886. The *Année sociologique*, the organ, of Durkheimean sociology, commenced a decade later, in 1896, while Max Weber assumed the editorship of the *Archiv für Sozialwissenschaft*, a more significant role than his university chair, in 1904. In Britain again - forgive me for the parochialism of these examples, but Britain is often thought, perhaps not without reason, to have been if anything somewhat backward in these matters - the Sociological Society and its journal the *Sociological Review*, and Englands first chair in Sociology, at L.S.E., all date from the first decade of this century. *The Economic Journal* had been publishing since 1890. The *American Journal of Sociology* was established in 1895. One could go on almost endlessly.

In one sense Durkheim's observation I referred to earlier about the ending of the heroic age spoke a kind of truth, which I have been trying to illustrate. but he also, by implication I think, exaggerated. It is, of course, easy to overstate the speed with which the newer social sciences, above all sociology, established themselves alongside the older academic posts and departments. Durkheim's chair at Bordeaux in 1885 was in education, not in sociology, and he was to wait twenty more years for this particular recognition. Weber, we recall, began his career in Freiburg as professor of economics, having been trained in jurisprudence. In England sociology had to fight long for establishment at Oxford and Cambridge, and arguably has not really done so even now, in marked contrast to the hospitality both extended at an early date to anthropology, in which chairs were established for Edward Tylor and James Frazer. But though these variations are interesting and I think, in the last case, significant in a way I want to return later, they are less important than the deeper sense in which Durkheim appeared to speak of Sociology - which he interpreted, of course, in a broad sense, including what elsewhere would have been called anthropology and psychology - as becoming an established science. The academic apparatus, which began to appear shortly afterwards, spoke surely more of hopes and aspirations than, as yet, of achievement. At the most what was beginning to emerge was some discernible methodological traditions within which scholars could work: Durkheimian, Marxist, Weberian and - above all, though of course not exclusively, in England - social evolutionist. But as yet it would certainly be misleading to speak as though aspiring social scientists shared a single and coherent paradigm; the long-cherished ambition to be the Newton or Darwin of the social sciences remained an ambition, and the *Année Sociologique*, the *Archiv für Sozialwissenschaft* and the *Sociological Review* testify as much to continuing conceptual and methodological diversity as to scientific aspiration.

To some extent the pretensions implied in the establishment of academic posts and journals only gave a positivistic gloss to the persistence of a variety of older traditions of scholarship and social theory to which the new masters and the new institutions were, sometimes with substantial modifications, the heirs: Comtean positivist in France, idealist and historicist in Germany, individualist, utilitarian and Malthusian in England - with, of course, a certain amount of cross-influence from country to country. Yet for all that I am inclined to risk what may seem like a paradox and say that in the nascent social sciences the third quarter of the nineteenth century, from the fifties to the eighties - the period, in fact, immediately preceding the one with which we are directly concerned - saw a greater coherence, a more marked coalescence, of different, largely national, intellectual traditions into something like a common set of concerns and concepts, than at any time since the mid-eighteenth century. I would be inclined, on the other hand, to see the period which concerns us here, of the very late nineteenth and early twentieth centuries, as exhibiting, in some degree a breaking-up of that consensus, while, inevitably, incorporating much from it. In case this sounds denigrating, I should also say that it seems to me that this partial fragmentation, if I am right to call it that, was also what gives much of its intellectual vitality to the period and constitutes its originality.

To try to justify this claim, and perhaps also to justify what may have seemed my over-long preamble, I want to return, in conclusion, but this time with specific reference to the social sciences, to what I spoke of earlier as an impatience with the constraints of historicism as a framework of thought. Because the consensus I want to attribute, with the obvious qualifications, to the third quarter of the century is essentially what I have loosely called historicism - a kind of partial confluence of Comtean notions of progress and neo-Darwinian or Spencerian social evolutionism, and German historicism. It seems often to have been easier than before, in a world of supposedly advancing co-operative scholarship, to ignore the wide differences, detaching the Comtean philosophy of history, as J.S. Mill tried to do, from the Religion of Humanity, or apparently subordinating the *a priori* necessities of Hegelian Idealism to the minutiae of historical scholarship. For there surely was some confluence. English ideas, under the sway of the concept of evolution, became more historically oriented, partially abandoning or at least muting the unhistorical influence of Benthamite individualism and classical political economy. German scholarship, in these years, while retaining its overriding preoccupation with history, often exhibited allegiances that seemed ostensibly positivist, even if Idealism often lay not far below the surface, compared with the metaphysical and Romantic influences which had shaped it earlier and the neo-Kantianism which became fashionable in the least years of the century. The Germans were still the schoolmasters of Europe, but they spoke in accents less alien to their contemporaries in other countries. One example of this is the close links, scholarly and sometimes personal, between German and English historians of the early middle-ages, working in the tradition of Jacob Grimm on *Die Germanische Rechtsgeschichte* (Burrow 1981: 119-26).

What constituted, above all, it seems to me, the scholarly common ground of the second half of the century was in fact an overall view of the history of European culture and institutions and a sense of its profound importance to the understanding of all social life. It was a view grounded in archaeology - the equivalent, for the humanities, of the palaeontological discoveries which had revolutionized the biological sciences - in the early systematic attempts of the mid-century to understand the prehistoric life of Europe revealed in Switzerland, in Denmark, in the Dordogne. From the eighteen-sixties onwards, too, there had been a renewed vigour in attempts to recover the intellectual and social prehistory of mankind from evidence provided by modern primitives. The study of early Teutonic land-law provided another ingredient, distinctive to Europe or at least to the 'Aryan' peoples. The emergence and break-up of feudalism provided the concluding chapters, as they had done in the work of the leading figures of the Scottish Enlightenment a century earlier and of Saint-Simon, Guizot and Marx later. Of course there were unsolved disputes, warring scholarly tribes, patriarchalists and matriarchalists, Aryanists and anthropologists, Romanists and Germanists, but these debates of the eighteen sixties and seventies only testify in a sense to how much of the scholarly stock was held in common, to a shared sense both of the possibility and the immense importance of such historical reconstruction, and to the shared possession, at least in its broad outlines, of a common story of the human and European past. It was a story which, though it incorporated elements of the speculations of the Enlightenment and the scholarship of the early nineteenth century, seemed to become more clearly etched, more profoundly detailed and more comprehensively illuminating from the middle years of the century onwards and it gives much of its distinctive intellectual character to the second half of the last century.

It was not simply that there was a notion of the importance of this historical scheme as a kind of matrix for all the sciences of man; there was even something like a consensus about its main concepts, methods and assumptions, about what constituted archaism, modernity and progress and how their manifestations were to be studied. In England it was referred to sometimes as the 'Comparative', sometimes the 'Historical' method; it was both. It is true, of course, that it is possible to see at least two distinct traditions of enquiry here, both stemming from the Enlightenment, but one primarily from France, the other from Scotland. There was the historical scheme which postulated stages of the human mind, from magic, through religion to science. It seems quite proper here to speak of an intellectual tradition, from Des Brosses, Concordet and Comte to Tylor, Frazer and Levy-Bruhl. The alternative scheme established most notably by Smith, Ferguson and John Millar, of the so-called four stages of hunting, pastoralism, agriculture and commerce, and incorporating also a kind of archetypal contrast between clan society and commercial society, with feudalism as the historical intermediary, focussed on law and institutions rather than stages of mind. The exact lines of connection to the eighteenth-century are often hard to draw, but of the two it was the latter categories which became the most profoundly influential in the nineteenth-century, providing the clue to the essence of modernity and progress, not only as the scientific mentality but as individualism, co-ordinated through the division of labour and the mechanism of the market. The distinctive character of progressive European society was given by the break-up of tribal society and the emergence of individual legal status and private property in land.

To try to illustrate the different ways in which this paradigm, as it seems tempting to call it, could be manipulated, I want very briefly to refer to three figures in the history of social sciences, from the eighteen sixties to the eighties, none of whom usually figures quite as prominently, I think, as they might do, in accounts of that history. I want to consider them because it seems to me that the contrasts as well as the similarities, between them and the later and greater figures of Durkheim and Max Weber are helpful in trying to see the achievements of the two latter in their historical context. In taking these three, of course, I do not pretend that the choice is not to some extent an arbitrary one. In particular I could, of course, have included Lewis Henry Morgan's *Ancient Society*, and its well-known influence on Engel's *The Origins of the Family, Private Property and the State* as well as its appearances in Marx's *Ethnological Notebooks*, or I could have considered Friedrich Bachofen or J.F. McLennan.

At the beginning of the eighteen sixties two books appeared, in England and France respectively, between which, though there was no direct influence, contemporaries immediately noted similarities: Henry Maine's *Ancient Law* (1861) and Fustel de Coulanges *La Cité Antique* (1861). Both dealt ostensibly with the early history of Rome, but in both also a general outline was clearly offered of the development of civilized society; both worked within a polar antithesis between the archaic and the modern, summed up in Maine's epigram that the movement of the progressive societies had been from status to contract, from a social order based on kin-right to one based on individuals related to each other by the rational intention and perception of mutual advantage embodied in the voluntary acceptance of a contract. *La Cité Antique*, as an account of the emergence of the classical city-state out of tribalism, is similarly conceived, though with the significant difference, characteristic of French social science, that it is presented more crucially in terms of religion. The ancient religion was the religion of the domestic hearth in which the unity of the family was embodied. Those who shared in the worship of the household gods were members of the kin-group. Fustel's story is in a sense a study in secularization, an account of the transformation of a society based on kinship, real or fictitious, bound together by a common cult, to one of men subject to a common law and participating in a common citizenship, and divided along lines of class.

My third figure is German. Ferdinand Tönnies in *Gemeinschaft und Gesellschaft*, published, of course, two decades after the two books I have just been considering, offers parallels to Maine and Fustel but with a significant difference. Maine's work is a history, though of a speculative and wide-ranging kind. Tönnies' contrast, expressed in his title, also, of course, refers to an historical transition: the archaic and familiar set against the modern and contractual. But Tönnies' work is significantly less directly historical than Maine's. Its polarizing of abstractions is more like what Weber was to familiarize in the literature of social science as the concept of an ideal type. Maine's is a work of scholarship. Tönnies', rather, of social science. Which brings me to the final part of what I wish to say.

Durkheim and Weber, the towering figures of the social sciences at the beginning of this century, seem to me to stand in relation to the large-scale historical schemes, from the archaic to the modern, that I have just been referring to, in a way that is both obvious and oblique. That the contrast, variously explored, between archaic and modern, between tribal solidarity and the division of labour, between tradition and bureaucratic rationality, between what Durkheim called 'mechanical' and 'organic' solidarity, is cardinal in their work seems undeniable, and in the broadest terms it was an inheritance from the previous period. Durkheim, it is worth remembering, had attended the École Normale Supérieure when Fustel was its Director, while Weber, trained as a lawyer, was heir to the intellectual traditions of German historical jurisprudence which Maine had been chiefly instrumental in introducing into English legal scholarship. Yet compared with their predecessors Durkheim and Weber seem to use these contrasts with a crucial flexibility. Durkheim does not write history at all, but uses the concepts it provides for conceptual explorations of modern society, on individualism, consensus, social solidarity. Weber does in a sense write history, and very large-scale history. Yet the concept of the scholar's focus of attention as being given by the value-relevance of its object, and the concept of the ideal type, as a deliberate conceptual artifice, applied to historical and empirical materials, were, we can surely say, emancipating. Emancipating, that is, from the intellectual matrix provided, in the previous generation, by social evolutionism or a comprehensive historicism, whichever it seems more appropriate to say.

Why, finally, might one wish, as I have done, to speak of emancipation rather than of, for example, fragmentation? To point the moral we can return to England. There is a wide agreement that English social theory or sociology in the early years of this century was less fertile, less conceptually inventive and less imaginative in the ways its attention was focussed, than social theorizing on the continent. If this was so then it seems to me that one way of describing it is to say that social theory in England remained generally locked into the social evolutionist or historical paradigm. In so far as there was eventually a break-out (rather than breakdown) it came first in anthropology and owed at least something to the influence of Durkheim. Hobhouse, in the chair of sociology at L.S.E., was still, in the middle of the nineteen twenties, working on yet a further revision of the general theory of social evolution, designed to replace the version of it first conceived by Herbert Spencer in the eighteen sixties (Hobhouse 1924). Such evolutionary schemes typically concluded with, and endorsed, a version of the present and future, valid for all progressive societies, seen in terms of individualism and rationality. In the earlier stages of history there was much to explain - non-rational belief, religion, various forms of non-rational social cohesion - so anthropology, as the study of backward peoples, flourished; not least, of course, because of its proclaimed utility to an imperial power. But about the present in England sociology seemed to find nothing theoretically inventive to say. In so far as modern life presented a problem calling for social investigation it was the problem of its marginal groups, a fact which fitted well with the close association, in its early days, between sociology and the organization of charity (Burrow 1966: 88-90). The most obviously disturbing, and even potentially dangerous marginal group was constituted by the extremity of poverty: the unrespectable poor, 'the submerged tenth', the *Lumpenproletariat*, above all, of London. Much attention was devoted to the question how far such poverty was attributable to innate or hereditary vice and weakness, how much to circumstances, and how far the circumstances might be remediable without the suspension of the need for self-discipline and self-help.

In thinking of the central practical problem for sociology as constituted by a group seen as marginal defined by its sufferings and even perceived as a kind of external threat to respectable society, sociology in England, it can be argued, marginalized itself on the maps of learning and in the academic world its status remained correspondingly low. The underlying assumption - and it was not altogether false - was the social and cultural homogeneity of respectable England, which could easily be tacitly read, in social evolutionist terms, as the essentially homogeneous and unproblematic character of 'modernity' as the end result of the evolutionary process. Those who did not fit, the unassimilated, were the poor of the great cities and the backward peoples of the empire, the subject-matter, respectively, of the sociologist and the anthropologist. Parallels, of course, were readily drawn. East of the strongholds of commercial rationality, the Bank and the Stock Exchange, one could speak of darkest London as of Darkest Africa. But in France and Germany modernity could more readily seem problematic, as a challenge to sociological curiosity and imagination. In both countries, for example, the army, insignificant in England, could seem at once the epitome of modern rationality and state power and a survival of the ancien régime. The Junker and the Burger, the republican secularist and the Catholic anti-Dreyfusard could seem distinct social species, yet not at all by reason of abject poverty. I spoke earlier of emancipation, as well as specialization, in speaking of the relations of Durkheim and Weber to historicism, because it seems that they inherited from it, but also adapted, concepts for analyzing the problematic, plural, disturbing character of modern societies. Durkheim's *conscience collective*, Weber's concept of charisma, did not simply lock the non-rational, as Spencerian social evolutionism did, into an isolated historical past. If the immediate result, for the social sciences, was less coherent than the historicist or evolutionist matrix, it was also more flexible, provocative and variously exploratory.

Bibliography

Burrow, J.W. (1966) *Evolution and Society*, Cambridge University Press, Cambridge, pp. 88-90.

Burrow, J.W. (1981) *A Liberal Descent.*

Caine, Sir Sydney (1963) *The History of the Foundation of The London School of Economics*, Bell & Co., London.

Clark, Terry Nichols (1973) *Prophets and Patrons. The French University and the Emergence of the Social Sciences.* Harvard University Press, Cambridge, Mass.

Hobhouse, Leonard Trelawney (1924) *Social Development: Its Nature and Conditions*, Holt, New York.

Hughes, H. Stuart (1959) *Consciousness and Society. The Reorientation of European Social Thought 1890-1930.* MacGibbon and Kee, London.

Lukes, Steven (1973) *Emile Durkheim: His Life and Work*, Allen Lane, London.

Pater, Walter (1873) *The Renaissance*, Macmillan & Co., London.

Rothblatt, Sheldon (1973) *Tradition and Change in English Liberal Educa-tion*, Faber and Faber, London.

Schiller, Friedrich (1958-60) *Briefe über die aesthetische Erziehung des Menschen, Sämliche Werke*, Hanser, München, Band III.

Shaw, George Bernard (1901) *Man and Superman.*

Wilde, Oscar (1891) The Critic as Artist. Richard Ellmann (ed.), *The Artist as Critic, Critical Writings of Oscar Wilde*, W.H. Allen, London 1970.

Philosophy vs. Sociology in Poland at the Beginning of the Century

Jerzy Szacki

The title of my paper is certainly too broad, for I do not intend to present the entirety of relations between philosophy and sociology or to describe each of them fully. This was a period of very animated intellectual life, a period in which completely different trends appeared from those about which I shall speak. My paper concerns only one dilemma of the thought of that time, namely, the dilemma between *creativity* and *causality*. This will be a small contribution to the subject of this conference: *The Humanities between Art and Science*. It seems to me, however, that this contribution concerns something of broader significance for the intellectual history of the turn of the century. Most briefly, I have in mind the debate on the nature and even the very possibility of the social sciences, a debate which was the consequence of the crisis (which continues to this day) of naturalism and evolutionism in the humanities. Some critics of the social sciences questioned not so much the particular approach to sociology dominant in the 19th century as the very idea of making social life the object of scientific research. To be sure, this did not mean invalidation of all sociological problems, but it did lead to a certain revolution in the division of intellectual labor: philosophy and history, from which sociology as a "natural science" had tried to liberate itself, once again returned to favor. At the same time, the assertion that the humanities, to which belongs the study of social phenomena, and the natural sciences are basically different, if not opposite, became a dogma.

As a spokesman for this point of view one can mention W. Dilthey, whose *Einleitung in die Geisteswissenschaften* contained a very passionate criticism of sociology as such. B. Croce took a similar position on sociology. For example, he wrote:

> On the other hand, if you strengthen the historical sense and, on the other, the philosophical sense (understanding philosophy as the science of ideals or values, that is, of the mind), so-

ciology, which lives a double life and gives us a little sophistic history or a bit of vulgar and imprecise philosophy, will vanish like fog in the sun (Quoted from Znaniecki, 1922: 211).

A similar liquidating criticism of sociology can be found later in Lukacs and Gramsci. I do not intend to review this literature here. I am simply stressing that the problems taken up in this paper are not an exclusive feature of Polish thought.

The heroes of my paper are three thinkers who were selected for the role which they played in Poland around the turn of the century and later and because their writings of this period have some relevance to the problem which interests me here. These three thinkers are Edward Abramowski (1868-1918), Stanislaw Brzozowski (1878-1911), and Florian Znaniecki (1882-1958), who is more well-known outside of Poland for his many years of activity in the U.S.A. They did not constitute a group. Perhaps they never even met. They probably were not aware that their views were similar on many points. The justification for comparing them is that they posed similar questions and provided somewhat similar answers. Since these questions were important and typical for the epoch, I shall present them here. The importance of these questions can be seen in the fact that they also appeared in more radical ideas, e.g. in Marxists situated within the boundaries of orthodoxy and basing themselves on the positivist theory of science. Ludwik Krzywicki (1859-1941), for example, also tried to present his views as "the philosophy of the creative act."

Scientific life in Poland before 1918 was organized only to a small extent, so the boundaries between the various disciplines of the humanities were very fluid in practice. That is why it is difficult to state briefly what these three authors dealt with. Abramowski cultivated philosophy, sociology, and psychology as well as the theory of socialism (as time went on he moved further and further away from orthodox Marxism). Brzozowski was primarily an outstanding literary critic and philosopher of culture, but he also called for a kind of socialist revolution. Znaniecki began as a poet, then for many years was a "pure" philosopher, and finally, under the influence of W.I. Thomas, became a sociologist, which he remained to the end of his relatively long life, not returning to most of his general philosophical questions. He became really well-known only as a sociologist. The interest in his works written before World War I is relatively recent. In this paper Znaniecki the sociologist is considered only marginally. He belongs to a different epoch and to a different intellectual formation.

All three of the thinkers discussed in this paper belong to the phase of Polish intellectual history which is most often called the "anti-positivist turning-point". The term modernism is also quite often used, though I avoid it here since it always had a limited meaning in English. In any case, it is probably unsuitable as a general name for what H. Stuart Hughes described as the reorientation of European social thought (I have in mind his book, *Consciousness and Society*). Speaking of the "anti-positivist turning-point", I have in mind three ideas which became widespread in Poland at the end of the 19th and beginning of the 20th century.

First, a new idea of philosophy which ceases to be treated as ancillary to and derivative from natural sciences but becomes an undertaking which is independent of the latter and the rival of the latter when it comes to the shaping of the world-view. The problem of philosophy no longer reduces to the *knowledge* of the world: the essential questions come to pertain to man's *living* in the world. Hence the very strong feeling of the defeat of the philosophy as it had been pursued earlier. Znaniecki would have said that the new philosophy had one of its sources in the growing disproportions between cognitive achievements, represented by natural sciences, and the needs of contemporary people who had to cope with moral, religious, aesthetic, etc., problems. It may be said that it is the sphere of *values*, and not that of *facts*, which has become the domain of philosophy.

Secondly, the anti-positivist breakthrough brought with it the collapse of the belief that the world forms an ordered and harmonious whole, and that the law of progress guarantees the future security and happiness of mankind. The world now appears rather as chaos in which human beings themselves introduce elements of order as far as they can do that. When seen in this perspective man turns out to be incomparably freer than he was in the positivist cosmos, where he was under the rule of laws of nature: he must now rely on himself and must make his own world which promises him nothing but uncertainty and struggle.

Thirdly, the intellectual atmosphere of the anti-positivist breakthrough included the belief that its epoch was one of a total *crisis of culture*: culture in its previous form has already exhausted its possibilities, and hence either a new, essentially different, culture is born or the social world must be overwhelmed with stagnation which will bring a new epoch of barbarity. When seen in that perspective philosophy could not be just a better or a worse theory of this or that, but had to offer a programme of a general cultural reconstruction, a promise, as Znaniecki put it, of "some new view of the world, more comprehensive, more productive, and more able to grow by creative additions" (Znaniecki 1919: 12).

Using K. Mannheim's expression, one can say that this new style of thinking is underpinned by a basic intention which is perhaps best formulated in the following statement by Brzozowski: "The only philosophy possible today is not the fatalism of progress, but the Promethean awareness of responsibility and struggle." I leave aside the question of the wider social context of this nascent "Promethean awareness", limiting myself to two observations whose development would require special research. First, the appearance of this new awareness in Poland coincides with the revival of the idea of armed struggle to regain national independence, which does not mean that the rhetoric of "action" always has direct political significance. Second, in many cases (i.a., the cases of Abramowski and Brzozowski) the fight to reorient philosophy is at the same time a fight to revive the idea of social revolution. The "fatalism of progress" is represented here not only by Comte and Spencer, but also by the Marxism of the Second International, to which Brzozowski opposes Marx interpreted in the spirit of activism if not voluntarism. These convergences are highly noteworthy, though I hardly wish to give any precise political meaning to the philosophical and literary slogans of that time.

The real subject of this paper is the question of social *science*, whose idea formed in the 19th century had to be challenged on the basis of the new style of thinking. For this idea consisted in reification of the social world, in approaching it as something external to human activity and the values created by it. The most radical was Brzozowski, who violently attacked the very idea of sociology, calling it "the most preposterous of fetishes":

> The very intention of a 'scientific' treatment of these matters is a symptom of the absolute disqualification of a researcher to understand anything of the real nature of society (Brzozowski 1937: 279).

The astonishing and unforgivable feature of sociology is that it "changes our life into something external to us", leads us into a state of illusion that the social world of our own creation is just as strange to our consciousness as the world of nature. He argues:

> The objective world, the world into which science leads us and which is accepted by naive minds as given, complete, and only discovered by man is entirely the creation of humanity (Brzozowski 1937: 454-455).

Brzozowski does not leave off with arguments or invectives against objectivism. He also tries to explain whose false consciousness this view of the social world is. Most briefly, he is inclined to see in objectivism the illusion of the intelligentsia, which, not participating in practical productive activity, tends to see the world as something alien to and independent of man. It is interesting that Brzozowski derives this inclination from romanticism, whose continuation in this respect is supposed to be positivism.

For Brzozowski the discredit of objectivism is at the same time the discredit of sociology using this completely false vision of reality. He declares for philosophy of a kind which does not pretend to be a science. What is really important in philosophy are values,

> the goals which a given philosopher of humanity has created or imposed. The world has no meaning for us until we put them into it. Every philosopher, if he was a philosopher, endowed the world with meaning, taught and forced people to experience this world as having such and not other meaning, hence to experience such and not another world. Thus every philosopher created his own world, in which he and those who followed him lived, and what made him a philosopher was based on this (...). *Philosophy is not a form of knowledge, it is something much deeper, it is a form of life, a form of incipient deed* (...). The man of science comes to know what is, what appears without his participation; the man of philosophy, the philosopher, prescribes, he wants to prescribe the direction of the world... (Brzozowski 1910: 64-69).

I do not go into the details of the philosophy of action promoted by Brzozowski, though it is unquestionably one of the most interesting and most radical reinterpretations of Marx, which is opposed here to Engels and all other "positivist" conceptions.

I began with Brzozowski not because he was the earliest chronologically, but because he seems to be the most resolute: there is simply no place for sociology here. Abramowski was much less radical. Brzozowski even called him a "positivist *au fond*" (Brzozowski 1936: 196), which does not make sense but shows the possibility of considerable differences between these authors.

I am convinced, though, that Abramowski's basic intention was the same as that of his younger colleague: to overcome the notion of social reality as something external to human activity and to introduce to its description the motif of free *creativity*, whose outcome is determined in no way by the initial state.

Abramowski writes about the basic contradiction between the *scientific method* and the *creative method*.

The basis of the first is causality, evolutionary determinism, and every historical fact, economic as well as moral, conscious and unconscious, is examined as necessary, determined by a whole series of previous phenomena; the second takes for its basis *free choice*, looks at historical facts as something which could have occurred or not, depending on the action of conscious human will, which determines only itself (Abramowski 1965: 72).

The starting point of these considerations is an analysis of socialism and, in particular, the status of socialist ideals. But Abramowski himself is aware that these considerations apply fully to every theory of society, since none of them has been able to cope with this dilemma, always falling into one-sidedness of one kind or another. Abramowski is troubled above all by the one-sidedness with which sociology adheres to the principles of scientific method, being unable to consider the role of "the creative factor, that is, deliberate choice", which dominates in art, ethics, and politics. Abramowski does not wish to eliminate sociology. He believes that it is possible to create a sociological theory which would overcome this contradiction of science and creativity. His "sociological phenomenalism", which I cannot discuss here, was an attempt to create such a theory by means of a homemade psychology. However, it is possible that he was never entirely certain whether sociology can introduce the factor of 'creativity' as something separate from needs and ideas drawn into the train of social causality or whether, on the contrary, if its fate is to reflect, as it has done till now, the changes spontaneously taking place under the impact of causes completely independent of human consciousness and will" (Abramowski 1980: 221).

Perhaps this was the "positivism" of which Brzozowski charged him, as I mentioned earlier.

From the very beginning Znaniecki struggles with the same problem of creativity and the unavoidable limitations of science, whose authority extends only to repeatable phenomena but is powerless when the explanation of new phenomena is required.

Science cannot entirely encompass what is absolutely creative, the process of formation of today's world. It cannot do so because it always looks for something unchangeable and repeatable besides changeability and creativity (Znaniecki 1912: 286).

Novelties cannot be explained causally; evolution - Znaniecki repeats after Bergson - is *creative*. We can leave out the cosmic aspect of Znaniecki's philosophy (which almost entirely absent in the *par excellence* social philosophies of the two writers discussed above) and limit ourselves to the conclusions that relate directly to sociology. The young Znaniecki - later a sociologist and enthusiastic supporter of sociology - strongly opposes what he calls the "sociological understanding of the world". The "very fact of creative development" speaks against it. I quote:

Experience may be given social character only in so far as it is considered in unchanged form - not in so far as it changes. Social phenomena do indeed impose themselves on the individual with an objectivity almost equal to naturalistic or ideal objectivity - but they impose themselves only when they are already finished, established, and the more constant they are, the greater their objectivity (...). Hence, if we assume that the world is entirely social, the feature of permanency is attributed to it without exception, since there is no place from which something new could come (...) The social world cannot enrich itself with absolutely new meanings on its own. So the sociological point of view can be applied at best to reality at a certain stage of development but not to development itself as creativity (Znaniecki 1912: 322-323).

Thus, in a certain sense, Znaniecki accepts the vision of the social world which Brzozowski rejected *in toto* and Abramowski thought (though I am unsure whether Znaniecki ever studied him), but refers to the real phenomenon of objectivization which takes place in social life, though at the beginning there is always someone's creative act. Sociology makes sense so long as we have to do with established social forms and actions which are routine. When society changes, bringing forth new forms and actions, sociology no longer has anything to say. Then we must rely on humanistic disciplines which deal with novelty and creativity.

Znaniecki's views on this point underwent certain changes which I do not wish to go into here. I limit myself to citing a view from the book *The Fall of Western Civilization* (1921), in which social reality is divided, so to speak, between sociology and the philosophy of culture. The former is a nomothetic science that investigates repeatable phenomena and is similar in this respect to the natural sciences, whereas the latter investigates "processes of creative development". These are two poles of social life which, in Znaniecki's opinion, never approach each other. As Brzozowski wrote: "What has a cause, is no longer creativity" (Brzozowski 1936: 358). The opposition between creativity and routine that is subject to the law of causality cannot be overcome. It is impossible to imagine that any science could describe one and the other simultaneously. Creativity is not the proper subject of scientific research; it is rather the domain of philosophy and art. Science, which attempts to discover causal relationships, focuses its entire attention on the world of things in which everything is finished and well-ordered. For this reason it has failed as the basis for a view of the world and source of practical guidelines. Science is useful in the sphere of material production, but it is useless as a teacher of life. Moreover, it can even be harmful in this role, pushing us in the direction of a view that everything is determined once and for all and that creativity is a 'psychological illusion'.

In the world of science - as Abramowski wrote - there is no place for a "deliberately acting party," since this world is governed by determinism: "there can be no party of astronomers forecasting planetary disturbances and, despite this, forced to act in this direction" (Abramowski 1965: 84). Thus the real meaning of the world is revealed to us not by science and sociology based on it, but by philosophy, which is concerned with practical reality, reality created by man which cannot be imagined in isolation from his activity. I leave aside here a certain important difference in their views: for Brzozowski this creativity is primarily work, while for Znaniecki it is intellectual activity that transforms the "ideal world", which the philosopher sees as primary in creative evolution. There are such differences on many other points, so a discussion of them would take much time. My only intention is to indicate the basic line of thinking.

Emphasis on the importance of creativity in the human world in all of the above cases is linked with highlighting the presence of the individual, the ego, in all social processes. There are many statements, especially in Brzozowski, to the effect that a disappearance of individuality would lead inexorably to stagnation and enslavement by the finished "world of things". Znaniecki links creativity with individuality no less strongly: "The stronger the tradition, the greater the social reality, and, on the contrary, the greater the role of creativity, the more dominant the individual nature of value" (Znaniecki 1913: 25). For this reason Znaniecki (and not only he) was rather critical of Durkheim, though he was one of the first Polish sociologists to become thoroughly familiar with the views of the French sociologist and to value them very highly - in so far as they could be regarded as referring to the "objectivized" side of social life. These points on individuality coincide with Polish literature of that time, which liked to use oppositions as poet and crowd, genius and "philistine".

The rhetoric of creativity and individuality which played such an important role in the works of the thinkers discussed, led either to a rejection of sociology as such, or to a considerable limitation of its rights, or, finally, to attempt to reconstruct it thoroughly by introducing what Abramowski called "individual elements". Interestingly enough, this revolt against sociology turned out to be incomplete in many respects, if not downright superficial. Naturally, the entire conceptual apparatus of positivist sociology, especially the notion of social laws, was completely destroyed (only temporarily in Znaniecki). At the same time, however, these writers hardly invalidate sociological *problems*. They remain sociologists more or less, though the social "organism" is now replaced by community of values or culture. In criticizing Durkheim, they assimilate certain elements of his way of thinking in spite of themselves. This is striking in Brzozowski, for example, whose attacks on sociology were especially vehement. First, he attempts to explain sociology *sociologically* as the ideology of a particular group of modern society, the intelligentsia. Second (which seems more important), Brzozowski comes out as a critic of traditional individualism. In his writings the thought recurs almost obsessively that "our I is always an outcome, a product; it forms behind our backs and was shaped for the most part before we came into the world" (Brzozowski 1937: 5). One more quote:

> Society as the field of human creativity, human initiative, a field absorbing the entire intellectual and emotional life of its members, remains unseen, as it were. All past thought is based on the *fiction* of the thinking individual directly communing with God, nature, the Absolute, etc. The entire emotional life of the individual is determined by social arrangements among which he lives: all of his limitations come from the same source (Brzozowski 1937: 467).

And so forth and so forth. Abramowski speaks up in the well-known dispute between Durkheim and Tarde and does not take the side of the latter, as might have been expected. He looks for some compromise formula and declares both sides to be partially right. Sociology should not claim the right to absorb psychology, but neither should it abdicate in favor of the latter. Social phenomena have a *psychological-objective* character; sociology should try to explain how physical phenomena are spiritualized and how mental phenomena are objectivized (Abramowski 1965: 90). Durkheim described the phenomenon of objectivization correctly. His mistake was to ignore the fact that what is subject to objectivization has its origins in the consciousness of the individual. The point is not to come out in favor of the individual or society, but to find a compromise that makes it possible to overcome this opposition.

Znaniecki in a similar way searches for an intermediate path between positions which he calls social psychologism and social ontologism, rejecting both of them equally. He rejects "ontologism", whose personification for him was Durkheim, for reasons which should be obvious in light of what I have already said. "Psychologism" is impossible to accept because the individual as such is a fiction. Culture, on the other hand, is a new extra-psychological reality that is not reducible to the actual experiences of the individual. There is no individual consciousness without social life. There is no individual as such. There are only historical individuals who belong to particular cultures and societies and are dependent on supra-individual schemes. Of course, Znaniecki does insist that a sociological description of the individual will never be adequate and comprehensive, since there will always be some irreducible remainder that is the source of creative initiative. Nonetheless, the truly creative individual is not an individual liberated from society. He is not a "philistine" who is satisfied with reaffirming existing social norms, but neither is he a "Gypsy" who rejects all social conventions. The concept of culture begins to play an increasing role in Znaniecki's thought in this connection. Culture is not only an indispensable regulator of the behaviour of individuals, but also an intermediary between them and nature. The mistake of naturalism was that it was unaware that nature is not given to us directly but only through culture. We become acquainted with nature already transformed through human activity and as members of a specific society who are unable to look, arrange phenomena, and understand differently than we have been taught.

> Our whole world (...) is permeated by culture, and we can no more imagine what was the world of our prehuman ancestors than we can imagine the fourth dimension. There is no way out of culture (Znaniecki 1919: 16).

That is why to Znaniecki studies of culture seem to be the harbinger of a new paradigm of all knowledge. As we can see, sociology, which was questioned at the starting point, regains its rights in the end.

To be sure, this is now a different sociology - one which tries to include "individual elements" and to examine individuals as active subjects of the social process. Its main idea seems to be well summed up by a methodological note to Thomas and Znaniecki's *The Polish Peasant in Europe and America*, the only work through which the conclusions of the discussions described here came to world sociology. We read there: "The cause of a social or individual phenomenon is never another social or individual phenomenon alone, but always a combination of a social and individual phenomenon" (Znaniecki 1969: 89). In this work we find an attempt to construct a sociology which could combine the search for social laws (Znaniecki returns to this "positivist" category) with an understanding of the role of the creative acts of individuals, an analysis of the "world of things" with an analysis of the "world of values". It is not my concern to investigate whether this attempt was successful. There is much evidence that even this reformed sociology was incapable of the philosophical assimilation of the idea of creativity. As I mentioned, Znaniecki believed that it was absolutely necessary to supplement this sociology with the philosophy of culture, whose task would be to investigate the creative aspects of social life.

Regardless of the final results of these inquiries, for the historian of sociology one very important conclusion comes from them. Understanding the development of this discipline is impossible as long as we do not transcend its boundaries, for what takes place in sociology in the narrow sense results largely from what happens in other disciplines and in intellectual history as a whole. What seems particularly important is the systematic investigation of the wide borderland between sociology and philosophy. Contrary to widely held views, the history of sociology is not only the history of a science gradually severing all links with philosophy. It is at the same time the history of a dialogue that continues to this day.

Bibliography

Abramowski, Edward (1965) *Zagadnienia socjalizmu* (Problems of Socialism), in: *Filozofia spoleczna*, Panstwowe Wydawnictwo Naukowe, Warszawa.

Abramowski, Edward (1980) *Metafizyka doswiadczalna* (Empirical Metaphysics), Panstwowe Wydawnictwo Naukowe, Warszawa.

Brzozowski, Stanislaw (1910) *Idee. Wstep do filozofii dojrzalosci-idziejowej* (Ideas. Introduction to a Philosophy of Historical Maturity), Ksiegarnia Polska B. Polonieckiego, Lwów.

Brzozowski, Stanislaw (1936) *Kultura i zycie* (Culture and Life), Wydawnictwo Instytutu Literackiego. Warszawa.

Brzozowski, Stanislaw (1937) *Legenda Mlodej Polski*, Wydawnictwo Instytutu Literackiego, Warszawa.

Znaniecki, Florian (1912) *Znaczenie rozwoju swiata i czlowieka* (The Significance of the Evolution of the World and of Man), in: *Swiat i Czlowiek*, vol. IV, Warszawa.

Znaniecki, Florian (1913) *Studia nad filozofia wartosci I, Elementy rzeczywistosci praktycznej* (Studies in the Philosophy of Values I. Elements of Practical Reality), Przeglad Filozoficzny, Warszawa.

Znaniecki, Florian (1919) *Cultural Reality*, University of Chicago Press, Chicago.

Znaniecki, Florian (1922) *Wstep do socjologii* (Introduction to Sociology), Gebethner i Wolff, Poznan.

Znaniecki, Florian (1969) *On Humanistic Sociology*. Selected Papers Edited and with an Introduction by Robert Bierstedt, The University of Chicago Press, Chicago.

Fin-de-siècle and the Mass Movements

Sven-Eric Liedman

In 1894, there appeared in Vienna and Leipzig a voluminous book entitled *Der Capitalismus fin de siècle* (Meyer 1894). The author's name was Rudolf Meyer. Who was he, and why did he write this book?

In vain you may look for his name in modern encyclopedias. But in the '90s, he was a famous man, who had once been the secretary of Bismarck and subsequently a vehement opponent of all Bismarckian politics. He represented an ideology called social conservatism or conservative socialism; Karl Marx labelled it "Polizeisozialismus."[1] His ideal was a great paternalist Fatherland, where the upper classes kindly but peremptorily took care of the lower ones. Capitalism was dangerous as it destroyed personal relationships and made economy the sole important factor in society. Indeed, for a short period capitalism was necessary in order to clear the ground for a relative social welfare, but according to Meyer this period had already come to an end.

Bismarck's reckless *Realpolitik* had made him deeply disappointed, but the young Kaiser's own adventurous policy was still more disastrous. Its consequent result would be a world war. Furthermore, it paved the way for the victory of the Social Democrats, who already had gained a lot from Bismarck's clumsy way of handling them. Only one force could now compete with the Social Democrats, namely a new - and not less dangerous - mass movement that took shape among catholics, revitalized under the guidance of the pope, Leo XIII.

According to the front page, Rudolf Meyer was the only author of *Der Capitalismus fin de siècle*. But at the end of the book, Meyer admits that he had had good help of a "young, very talented writer, Herr Doktor Paul Ernst." (Ernst 1931: 248-49; Meyer 1894: "Postscript").

This collaboration is worthy of some consideration. Much later when he was well-known as an author of dramas in neoclassical

---

[1] In a letter to L. Kugelmann 18.5.1874, Marx calls Meyer "Polizei-Socialist, faiseur, literarischer Sudler" (Marx Engels Werke (MEW) vol. 33: 627).

style, Paul Ernst has shed some light over his assistance to the ageing Rudolf Meyer. The young Paul Ernst was just as much on the left side of the political arena as Meyer was on the right. Ernst had been one of the leading members of "Die Jungen," the leftist group inside German Social Democracy which had played an important part at the famous Erfurt Congress in 1891, where the new, more Marxist inspired party programme had been accepted and where some of "Die Jungen," albeit not Ernst himself, had been excluded because of leftist deviations.[1]

Ernst was an academic. He had studied Marx. He had listened to the lectures of the most famous Prussian professors in Berlin - and turned away from them in disdain. "To go from Schmoller to Marx is heaven!" he exclaimed.[2] On his way from the university to his home, he met the new proletarian crowds in the streets. The view evoked a mixed sense of veneration, compassion and vague hope in him. One of his favourites was Leo Tolstoy. He made a famous quotation from Tolstoy his own: "The masses, the immense masses have been and still are the carriers of the answers to the vital questions."

As many other young German students and artists around 1890, Paul Ernst turned out to be a Marxist and socialist. In 1890, Bismarck resigned and the law which forbade socialist political activities was abolished. Ernst and his comrades found themselves free to propagate their convictions. But the time was short, they felt; revolution had to be carried out immediately or at least to-morrow. They felt contempt for older Social Democrats such as August Bebel or Wilhelm Liebknecht who trusted the general elections as a means for coming into power. In turn, they hoped for assistance from Friederich Engels, the old socialist guru in London. But in his response to them, Engels was bitingly discouraging. As we shall soon see, Engels himself was much more optimistic about the possibilities of a socialist revolution to accomplish out so to say from the inside of the German institutions.

After Erfurt, "Die Jungen" ceased to be a real group. Some of its members turned out to be anarchists, other loyal Social Democrats. When meeting the old conservative "Polizeisozialist" Rudolf Meyer, Ernst was ideologically and politically homeless.

---

[1] The fate of Die Jungen has been newly treated by two authors (Müller 1975; Wienand 1976: 208-41).

[2] "Ich kam von Schmoller zu Marx und kam wie in den Himmel" (Ernst 1931: 167).

In some respects, his ideas may have been the real opposite of Meyer's. In others, they converged.

The problem of the masses represented the focal point to both Meyer and Ernst. Here, we can already discern the contours of the stereotypes of the "fin de siècle" context: the lonely individual, "der Einzige", sometimes the superman, in front of the masses. This is the paradigm of Nietzsche and not only Nietzsche.

In the stereotype, however, the superman and the masses are depicted in a very abstract, lifeless way. It has not only to do with our lack of imagination but also with our own concrete situation. To-day in most European countries as in Northern America, the concept of 'mass' is without smell and taste. Our streets are over-crowded not by men but by stinking, lifeless cars. The masses can only occasionally constitute themselves as an electorate, as a lot of isolated people looking at a popular TV-programme, or, most concretely, as the spectators of a foot-ball match or as the partici-pants of a, normally well-arranged, political demonstration. Fur-thermore, the industrial development means that even most big industries tend to be relatively sparsely populated places.

The masses of the fin-de-siècle Europe, as the masses of modern China, India or Iran, were physically present in all big cities, most of which were growing in such a tremendous way that the vast majority of their inhabitants were not born in the city. To Paul Ernst, the masses meant the crowds of workers on their way home from job. To Rudolf Meyer, it was the thousands of participants of the Social Democrat demonstrations but also the emerging new masses that the Catholics and the nationalists set in motion.

In the '90s, there were different kind of masses - leftist or rightist, socialist or nationalist, rabulist or pious. The year 1890, which is still a year marking the start of a new epoch in many textbooks in modern history, was originally chosen as such a memorable year not because of the fin-de-siècle pessimism, deca-dence or anti-positivism, but as the very year when Bismarck was overthrown, the anti-socialist laws were abolished and socialist parties won unprecedented electoral victories in many European countries. 1890 also means the definite start of the May Day demonstrations as an international manifestation of working-class aspirations and socialist politics. The demonstrations deeply impressed friends and foes. "At last, England's working classes set in motion," Engels reported enthusiastically from London.[1] "The mob in the streets!" the adversaries murmured in London, Paris and Berlin.

In hindsight, when we have the final result of 1914, it is easy to say that these so seemingly committed and dangerous socialist crowds already were on their way as innocent victims to the trenches of the First World war. However, we must remember that this result in all its cruel reality was not only not possible to foresee from the horizon of 1890, but that it also must be put into question if the process in some hidden way really was inevitable. In any case, the socialists in the streets were the cause of fear and enthusiasm in those years.

As already touched upon, there were also other masses than the socialist ones. First of all, nationalism itself came into existence as a mass movement. One may also say, that nationalism or patriotism in those very days definitely became rightist and that it then had to form itself as opposite to all leftist movements. Still in the '80s, "General Revenge," Georges Boulanger, in France, who headed a short-lived mass movement of nationalist character, in his campaign could collect everything from old left-wing revolutionaries to right-wing legitimists.

The political right as a mass movement is one of the new, ominous and definitely never forecasted emergences of the fin-de-siècle period.

There was also a third kind of crowd which, although physically absent, played an important part in the conceptions of the '90s. It was the crowds - no, the hordes of Russia and China, who threatened all European culture. With the assistance of Paul Ernst, Rudolf Meyer wrote in his book about the end of capitalism, I quote:

We are facing a World War. The struggle against Russia is a continuation of the ancient struggles between east and west... It is the struggle of culture against barbary, the struggle of progressiveness and freedom as against Asiatic despotism and stagnated servitude (Meyer 1894: 245).

In the '90s, an impending World War belonged to the frequent topics of discussion. Again, the ideas of the masses played an important part. Friederich Engels and many other leading socialists

---

[1] "Jetzt endlich scheint das englische Proletariat in Masse in die Bewegung zu kommen", Engels writes in a letter to F. A. Sorge 30.4.1890 (MEW Bd. 37: 398)

saw the socialist mass movements as the only guarantee for peace. Engels wrote in 1891 some very optimistic lines for a French socialist almanac about the probable development in Germany. The figures from the general elections demonstrated the unbroken success of the Social Democrats. If the trend would go on - and on this point, Engels did not hesitate, as according to his opinion the new generations were predominantly socialist - the Socialist Democrats would in the long run gain absolute majority in the German parliament.

There was however, a means to take over power that was more secure and more rapid - the army. Engels says:

> The main strength of German Social Democracy is not the number of its voters. In Germany, the elector is 25 years old, the soldier only 20. And as it is the young generation that first and foremost delivers our voters, it follows that the German army steadily becomes more and more permeated by socialism. To-day, we have one soldier out of five, in a few years we shall have one of three, and around 1900 the army, formerly the most Prussian element in the country, will become predominantly socialist (Engels MEW bd. 22: 251).[1]

According to Engels, this not only paved the way to political power but also meant a future guarantee against a World War. However, he admitted that the outbreak of such a disastrous war depended not only on Germany; an alliance between Russia and France against Germany could make a war inevitable. Hence, one of the main tasks of the newly founded II. International was to bring about a close cooperation among all socialist parties in order to fight against all martial tendencies in every country.

As we know, this was a task that the II. International really fulfilled, although in the long run without succeeding in hampering the catastrophe of 1914.

To-day, Engels' forecasts about the future success of the Social Democrats inside both parliament and army may seem naïve. However, his mistake is of the same kind as most of us - and even many modern professional futurologists - still make. He is extrapolating a strong tendency, not calculating that such a victorious tendency must awaken its opponents to resistance and, through the mere burden of success, itself tends to be more

---

[1] I am grateful to a graduate student, Amanda Poralta, who has called my attention to this quotation.

undecided or even split up. On this point, he is just as undialec-
tical as most of us. In the '90s, his kind of reasoning was far from
unusual. To his adversaries, the intrusion of Social Democrat influence in parliament and army caused many nightmares.

For a while, we may turn back to the masses. The conception of
the masses was intimately associated with the conception of the
individual. The individual might be the leader, organically
united with the movement as August Bebel in the German Social Democracy. He might be a genius in the service of mankind -
as the great scientist or artist. But he might also be the one who in
contempt or loathing turns away from the crowds, convinced of
his own superiority and the vanity of everything.

The first thing we must remember is that the conception of
'the masses' is inseparable from the conception of the great individual.

It seems as if this in some respect rather trivial but still absolutely fundamental connection is often overseen in many historical textbooks and even in more fundamental historical investigations. We are still victims of a type of periodization, according
to which each epoch is seen as characterized by a set of dominating trends or tendencies and ideas - and the corresponding lack of
its opposites. Thence, it is the pessimistic, lonely, degenerated
hero who is the emblem of the fin-de-siècle, whereas his necessary counterpart, the masses outside his windows, without which
his own character is impossible, are forgotten.

In this respect, I have closely studied two eminent books in intellectual history, H. Stuart Hughes' *Consciousness and Society:
The Reorientation of European Social Thought 1890-1930*, and
Fritz K. Ringer's *The decline of the German Mandarins: The
German Academic Community, 1890-1933*, both of which have
their starting point in the ominous year 1890. I have found that
both Ringer and Hughes presuppose, that our period has a few
dominant, in some respects rather clear-cut intellectual traits in
common. In his impressive investigation of the German academic intelligentsia, Ringer says:

> Sometime around 1890, German academics began to express
> misgivings about the current condition of German learning
> and German cultural life more generally (Ringer 1969: 253).

This is the famous fin-de-siècle pessimism, here transposed to
the not very exotic level of German professors, who according to
Ringer were afraid of soon losing their favourized social position

and status in front of industrialization and the success of the engineer.

Let us, however, take the issue of periodization from a more critical point of view and ask: What reasons do we have for choosing the year of 1890 as a crucial one in intellectual history? It is not unusual for intellectual history to take over periodization from other kinds of history - political, social, or economic - and accordingly for new chapters in the textbooks of intellectual history to begin with 1789, or with the industrial revolution, or the age of capitalism, or suchlike. Such divisions may be fairly good, if those changes also had immediate and substantial effects on intellectual affairs. The year 1890, originally a demarcation line in political history, may (as I actually suppose) also be a convenient starting point in intellectual history if something new in modes of thought and in intellectual behaviour can be said to have really started roughly at this time.

In keeping with the perspective used in this article, I want to qualify this position. The novelty ought to be visible in new topics under discussion and new controversies emerging. I am sure that Ringer is right to see the normally unnoticed intellectual habits and the unconscious or semiconscious preferences as the real foundation of explicit ideas. Intellectual history must, however, take its beginning in what it has to explain, that is, in expressed ideas, and from them it must search for its roots. It is improbable that new mental habits can be dated with any precision, the changes being so general, so widespread, and so difficult to localize. The changes that we notice in intellectual history are changes in topics under debate and visible attitudes to former topics. Every age, every milieu, has literally its own universe of discourse. A fruitful periodization of intellectual history points at substantial shifts in such a universe of discourse.

To characterize an intellectual milieu, then, will always mean pointing out what is under debate within it, recognizing how a stand-point on one issue is influenced by stand-points on another. However, it also means finding out what is hidden, what is forgotten, what the participants are unable to see - and why. The mapping of what is being debated necessarily leads to the more fundamental considerations of the inner logic in intellectual history and its roots in human affairs. It is, however, incorrect to treat intellectual history as only epiphenomenal and as reducible to other kinds of history.

Consequently, the 1890s form a genuinely new period in intellectual history only if this era is marked by a substantial change as to the universe of discourse. Characterization of a given universe of discourse is not fundamentally concerned with identifying the sympathies and antipathies of those who lived during the particular period. Rather, we must primarily characterize the issues under discussion. The anti-positivists of the nineties presuppose the positivists, just as anti-industrialists are counteracted by the apologists of industry. If there was something new about positivism and its opponents, this would mean that the whole question of positivism (and of science, of progress) acquired new meanings, new consequence, new contexts. The difference between the '80s and the '90s cannot be seen merely as the step from positivism to anti-positivism any more than the difference between German and English intellectual milieus can be seen as the difference between anti-industrialism and pro-industrialism. To my mind, the shift seems to be more complicated. The earlier anti-positivism - *positivism* here being used in its original, robust, Comtean meaning- was substantially an older view than its adversary, opposing science, technology, enlightenment, and other modernities in the name of tradition, religion, and privilege. The new critique of scientific optimism and rationalism had in a way integrated this positivism and made it its own. Nietzsche, the forerunner, presupposed many extreme ideas of social Darwinism and scientism. Ernst Mach rejected system-building positivism only to construct his own neo-positivism; and even those people who became irrationalists and mystics had lost this kind of innocence in relation to scientism and modernism, and innocence that characterized much of earlier anti-positivism.

The case of positivism and anti-positivism, however, is a rather vague one. To make the point clear, we must choose a more distinct phenomenon, and I do not hesitate here to raise the case of Marxism as being much more illustrative, at least in the German milieu.

Marx, seen both as the author of *Capital* and as the leading theoretician of international socialism, was of course not unknown to German intellectuals before the '90s. The conservatives and the liberals hated him and wrote not a little about him, although they normally saw his theories as nothing but emanations of his horrible revolutionary mind. Among his adherents and followers, very few in Germany were academics, the vast majority being unlearned people.

The situation changed drastically around 1890. A lot of young people at the universities became Marxists. Many of them had literary and artistic interests of a naturalistic trend. Not a few became members of the Social Democratic party, and as such they formed together with other young people the leftist, partly anarchistic wing called Die Jungen. As we have already seen, the success of Die Jungen inside the party ended abruptly with the congress at Erfurt.

The young, originally revolutionary students were not alone at the German universities - especially Berlin - in taking Marx seriously during the '90s. A vast, truly academic literature concerning Marx and Marxist theories appeared in the decade before and some years after the turn of the century. The change is to be seen not only in the followers that Marx was gaining at the universities but also in the fact that all adherents, adversaries, and even lukewarm historians and philosophers, such as Max Weber, were having to take Marxist ideas seriously.

Marxism is, of course, only one of many possible examples, the case of Nietzsche being another, the new type of critique of Darwinian evolutionary theory a third, the emerging crisis of the Newtonian world picture a fourth. The enumeration could continue. However, my intention is not fully to develop any of those examples.[1]

Instead, let us return to the case from which we started and which I have not forgotten: the strange cooperation of Meyer and Ernst. How can we understand that a right wing opponent of capitalism could join forces in a common project with a left wing opponent of the same capitalism?

Originally, it was in dealing with the problems in modern agriculture that Ernst and Meyer at first found that they agreed. Both rejected large-scale production in this field, irrespective of whether the regime was capitalist or socialist. In all, even Ernst always had a very ambivalent attitude to technical and industrial development. Here, he was far from alone among the young socialists and revolutionaries at the turn of the century. Many of them were at least undecided about the benefits of the enormous changes that took place.

When still an undoubting Social Democrat, Ernst's sympathy for proletarians and socialists had dominated his political ideology. Gradually, his attitude changed. He turned out to be an anti-modernist in all respects. It is not by mere chance that he developed into a classicist in his dramatic writings.

---

[1] See further my article "Mandarins and non-mandarins" with a response from Prof. Ringer and a final remark from myself (Liedmann 1986: 119-68).

Let us summarize. Already when Ernst was a revolutionary social democrat, there were central issues on which he and Meyer agreed. Their disagreement was, however, still fundamental. Ernst strived for a society where the proletarians and their political party had the ultimate power, Meyer on the contrary for a society where the same working-class was an object of the benevolent but resolute guidance of the traditional upper classes.

However, Ernst's attitude changed. He repudiated not only Social Democracy but all radical and political change in a leveling direction. Now, he and Meyer could agree in what was most important to both of them. They had their anti-capitalism and their negative attitude to large-scale production in common.

Our example demonstrates that every ideology and every *Zeitgeist* as "fin-de-siècle" (or enlightenment, or romanticism) has to be seen not only as loose and vague but first of all as multidimensional and full of inner contradictions and oppositions. Fin-de-siècle is not only the time of pessimism and worship of heroes - it is also a period when the masses and the *conception* of masses played an important part.

Bibliography

Ernst, Paul (1931) *Jünglingsjahre*, München.

Liedman, Sven-Eric (1986) Mandarins and non-mandarins, in: *Comparative Studies in Society and History* 28,119-68.

Marx /Engels *Werke* (MEW), Bd. 22, 33, 37.

Meyer, Rudolf (1894) *Der Capitalismus fin-de-siècle.* Wien and Leipzig.

Müller, Dirk H. (1975) Idealismus und Revolution. Zur Opposition der Jungen gegen den Sozialdemokratischen Parteivorstand 1890 bis 1894, *Beihefte zur Internationalen wissenschaftlichen Korrespondenz zur Geschichte der deutschen Arbeitsbewegung* 3, Berlin.

Ringer, Fritz K. (1969) *The Decline of the German Mandarins: The German Academic Community 1890-1933*, Mass, Cambridge.

Wienand, P. (1976) "Revoluzzer und Revisionisten. 'Die Jungen' in der Sozialdemokratie vor der Jahrhundertwende", *Politische Vierteljahrschrift*, 17: 2, 208-41.

Towards a History of the Melting Pot, or Why There is a Chicago School of Sociology but not a Detroit School

Jonathan Matthew Schwartz

Introduction

The "melting pot" is one of the most vivid and deceptive metaphors of American civilization. To imagine political society as a huge, *living* creature - as did several of the classical political theorists - made much more sense. The "melting pot" was a *new* idea, and that made a difference. Israel Zangwill, an English author who himself never entered the American melting pot, was the person who gave the image its first dramatic form. He wrote a melo-drama, "The Melting Pot," which began a popular run in 1908 (Higham 1975: 124-5).

The melting pot assumed at the opening of the twentieth century a significance for American culture and society that replaced the image of the 'frontier'. The melting pot was essentially urban, and of course, it was industrial. It seemed to fit the reality of intensive immigration, and if the Statue of Liberty greeted the hordes of immigrants at the mouth of the Hudson River, the melting pot represented the typical experience of the years and generations to come. The promise - or was it a threat? - of Americanization took place in that melting pot. In the crucible many metals turned into one alloy: *e pluribus unum!*

There have been critics of the melting pot symbolism, as well as staunch advocates. The terse statement of Professors Moynihan and Glazer: "The point about the melting pot is that it did not happen," heralds a discovery of ethnic resistance to assimilation into American society (Glazer and Moynihan 1963: 196).

This present paper is likewise cast in the spirit of dissent, but I shall emphasize the power of the melting pot as a symbol, and even as a ritual, in Americanization theory and practise. We suspend for a time the question of the truth or falsehood of the sym-

Jonathan Matthew Schwartz

bol. To reveal the "symbolic capital" of the melting pot, I shall contrast the research paradigm of the Chicago School with the applied sociology of Henry Ford in Detroit. The paper should help explain why a *Detroit* School of social science never had a chance to develop.

Chicago and Berlin at the Turn of the Century

The emergence of the Chicago School represented a sustained, collective effort by a private university that "first opened its pseudo-Gothic doors in 1892" (Coser, L. 1971: 344.)

The University of Chicago was generously endowed by John D. Rockefeller, whose style of capitalist patronage was opposite to that of Henry Ford. Among the innovations of the University of Chicago was that it created the world's first department of sociology (Vidich and Lyman 1985: 178.)

The University and its Sociology Department opened in the same year with a zealous spirit of reform. The philosophers, theologians, and social scientists in Chicago were able to carry out empirical and practical research. They were also some of America's best spokesmen for pragmatism, and nearly every leading figure in the Chicago School of Sociology had had some training in a German university. The transmission of ideas from Berlin to Chicago is a notable facet of the Chicago School. The salient feature of the Chicago School, however, was its naturalism. Its members entered directly into "the natural areas" of the city to study and understand its wild life[1]. In close

---

[1] To indicate the similarity of method and temperament in the social work movement and the social science at the University of Chicago, one could consult the Hull-House Maps and Papers: *A Presentation of Nationalities and Wages in a Congested District of Chicago, together with Comments and Essays on Problems growing out of Social Conditions*, 1895. Settlement House studies and Chicago Sociology both were active in empirical, documentary, and close-range field work. See Hortense Powdermaker, "Field-Work," *International Encyclopedia of the Social Sciences* 5-6 - (1968) 423.

"Social Disorganization" was a typical judgement, both descriptive and explanatory, of urban society. The term appears often in the writings of W.I. Thomas and Florian Znaniecki. However, in their work on Polish immigrants, (1918-20). Thomas and Znaniecki note the persistence of "peasant community" in the New World setting. "Disorganization" and "uprootedness" were by no means total. "Community" was, for the Chicago School, a key concept in sociological understanding. See Martin Bulmer, *The Chicago School of Sociology* (Chicago, 1984).

association with social *workers*, the social *scientists* addressed themselves to the problems of urban society: juvenile delinquency, slum housing, criminality, drug addiction, all of which were seen as symptoms of profound social disorganization.

As mentioned, the Chicago sociologists had studied in German universities, Berlin being the favored milieu. An expression of their pragmatism was the Chicago sociologists' adeptness at transplanting *just* those insights from Berlin which could meet their needs in the New World. Robert Park, who became the dean of Chicago Sociology in the 1920s, had attended lectures of Georg Simmel, at the University of Berlin in 1900. That was to be Park's only formal training in sociology. Park and Burgess translated and published in English two essays of Simmel, notably, "The Stranger," and "The Metropolis and Mental Life." A small bit of inspiration worked wonders in Chicago!

A similar hasty exchange occurred with Max Weber's brief visit to the eastern United States in 1904. After several years of emotional paralysis, Weber spent a few months traveling the United States. This loosened up Weber's sociological imagination. He seemed to thrive outside of stuffy Wilhelmian Prussia. He liked the long train rides in America, listening to stories of traveling salesmen in Pullman cars, while Robert Park sought out the seminal minds of *Geisteswissenschaft* in Berlin. Weber's observations of "maddening" Chicago are more in the spirit of expressionism than of sociological naturalism. The city was, according to Max Weber, "like a man whose skin has been peeled off and whose entrails one sees at work" (Gerth and Mills 1958: 16). The circulation of experience between Chicago and Berlin during the first decade of our present century would indeed be a worthy study in itself.

Before shifting my attention to Detroit - which was left out of that interesting sociological *kula* - I would like to sketch in broad lines some characteristic themes in the Chicago School paradigm.

The first studies of the Chicago sociologists were informed by a journalist whose formal training in sociology was minimal. W.I. Thomas and Florian Znaniecki, who collaborated on the monumental study of *The Polish Peasant in Europe and America* during the 1910s, drew upon documentary materials of emigrants and their communities, like letters between family members in the Old Country and in the New World. Thomas and Znaniecki also consulted newspapers for significant "letters to the editor" as well as the files of social workers and priests. Their aim was to create a series of collective biographies, "life histories", of common people that could serve as a valid basis for generalization. Portraiture that maintained actual likenesses more than abstract typologies was the trademark of the Chicago School. Social science - not unlike naturalist novels and settlement house reports - had to keep its eyes and ears open to the life of the "community". If Chicago Sociology may have had its initial impetus in theology and social philosophy, it soon turned its attention to the "here and now". Thomas and Znaniecki wrote of their method:

> And thus the essential criterion of social science as against social philosophy is the direct dependence of its generalizations on new discoveries and new happenings (Thomas and Znaniecki 63).

"Community" was - and still remains - a root concept in Chicago Sociology possessing empirical *and* theoretical vitality. "Community" (unlike *Gemeinschaft*) was not the exclusive possession of the small, rural society. Tönnies' juxtaposition of traditional *Gemeinschaft* and modern *Gesellschaft* underwent considerable modification, if not metamorphosis, when it entered the Chicago Sociology. In study after study, the Chicago sociologists observed the survival and rejuvenation of "community" in the urban metropolis.

More fitting than the dichotomy between "Community" and "Society" *à la* Tönnies was the Chicago School's flexible concept of "acculturation." The neighborhood and the ethnic community were active forces in the process of acculturation. Tönnies' nostalgia for the rural hearth and home which is epitomized in his *Gemeinschaft* played no significant role in the Chicago social science. Curiously, the nostalgia for rural conduct and culture will reappear in the principles of the "Flivver King", Henry Ford.

For Robert Park and his students, Chicago was a "social laboratory" where one could observe the human field much in the manner of anthropologists among small "folk communities." Park was also struck by what he termed "the ecology" of human communities in the city. He intended with the analogy to show how different immigrants' groups and social strata came to occupy urban space. "Ecology", a term coined by the German evolutionary biologist, Ernst Haeckel, in the 1870s, was one of the major items in the assimilation of European scientific ideas by the Chicago School (Bates 1953, Hannerz 1980: 26-30).

The Chicago School was pre-eminantly and explicitly "humanistic" in its method, its judgements, and its remedies for social ills. As with Georg Simmel, the Chicago sociologists were tempted by aesthetic concerns, by literary imagination, and some of their field-work resulted in sensitive, vivid descriptions that verged on naturalist novels. Novels, we know, could also be set in Chicago! It is not too long a jump from Theodore Dreiser's *Sister Carrie* (1900) to Zorbaugh's *The Gold Coast and the Slum* (1929), a description by a sociologist of Chicago's North Side. If one had to choose *the site* for primordial, inter-disciplinary studies, it would have to be the University of Chicago in the early years of our century. It was a university milieu where social scientists could rub elbows with poets and novelists and draw inspiration (and desperation) from what was going on a few blocks from the "Midway".

The Chicago School "aesthetic" borrowed no more than it needed from Georg Simmel - only those essays which regarded the urban world as a stimulus for individuality and eclectic taste. The Chicago sociologists by-passed the formalism of Simmel's social thought, though in George Herbert Mead's interactionist social psychology, Berlin and Chicago seemed to converge independently.

Both Berlin and Chicago were cities that sprang into existence and prominence in the final decades of the nineteenth century. By European, and even by German standards, Berlin was not an "old" city. Köln and Frankfurt were more deserving for historicist study. *Modern urbanism*, however, seemed to thrive best in cities which grew up overnight and where the shock of the "modern" gave fascination as well as anxiety. An unbroken tradition of modern urbanism can be discerned in Chicago literature from Dreiser to Bellow. The spirit of urbanism was never better described than in Louis Wirth's essay of 1938, "Urbanism as a Way of Life."

I cite the passage which utilizes the image of the melting pot in an unorthodox manner:

The city has thus historically been the melting-pot of races, peoples, and cultures, and a most favourable breeding ground of new biological and cultural hybrids. It has not only tolerated but rewarded individual differences. It has brought together people from the ends of the earth *because* they are different and thus useful to one another, rather than because they are homogeneous and like-minded (Wirth 1938: 10).

Henry Ford's Melting Pot in Detroit

The cultural pluralism that blossomed in Wirth's Chicago was nipped in the bud in Henry Ford's Detroit. The "Sociology" which Ford applied in the management of his labor force learned nothing from the Chicago sociology. Perhaps the neglect was intentional. True enough, the Chicago School looked primarily at the sphere of "reproduction", as Marxian theory would put it. The Chicago sociologists were more interested in the kitchen stoves than in the blast furnaces. Industry recedes from their urbanism. Ford's sociology, on the other hand, could hardly separate the factory from the home. Scientific study and management could be brought to bear on both spheres, though clearly the needs of industry determined the conduct of the household. Henry Ford's melting pot was the assembly line. Ford's sociology, therefore, took its inspiration from Frederick W. Taylor's "Scientific Management" (1911) and from uplifting Americanizers such as Peter Roberts (Roberts 1912). The two authors gave Ford an ample supply of sociological method and theory. There was no need to look to the professors in Chicago.

Detroit existed as a geometric grid before it received its population. The plan for Detroit was executed in the 1870s. In 1880 the population of Detroit was a modest 116,000 (Zunz 1977: 109). Chicago, 280 miles to the west, had already approached the one-million mark. By 1900, Detroit's population reached 286,000. It is only very recently that Detroits' ethnic "mapping" has been examined for this early period. Zunz (1977: 109) by studying demographic data, has been able to identify ethnic settlements on the grid of Detroit. Social workers in the settlement house movement of the 1880s were already aware of ethnic communities in Chicago, their housing and working conditions. The situation in Detroit was completely different. Sociologically Detroit did not exist, until the Ford Motor Company introduced the term in 1914.

The real acceleration of Detroit's demographic and sociological growth came with the mass production of automobiles in the first three decades of the twentieth century. By 1920 Detroit had a population of nearly one million, four times that of the year 1900.

The vacant grid was rapidly filling out. Contemporary descriptions of the city's residential districts emphasize the haphazard building. Factory workers, doing shift work, took turns with the same bed. The rational plan of the spoke-like avenues running from the city center to the suburbs bore little resemblance to the chaotic reality of the town. Detroit looked and felt like a frontier: rooming houses, saloons (before Prohibition) and factories. The railroads in Detroit served the factories, not the working population.

Henry Ford did not build his auto plants inside the City of Detroit. His first assembly line was in Highland Park, a plant that opened in 1909 on property that had been used as a race track. In the 1920s Ford shifted manufacture to the new, gigantic River Rouge works in Dearborn, again outside the city limits of Detroit, where most of his employees lived. Ford reckoned that smaller city halls were easier to control than expansive Detroit.

Ford's Sociological Department was an attempt to extend "scientific management" to the sphere of reproduction. "I am more a manufacturer of men than of automobiles," Ford once remarked with his usual gift of self-advertisement. The high rate of turn-over of the labor force at the Ford Highland park plant was management's biggest problem. It apparently was not easy to manufacture men! The speed-up of the assembly lines, the continuous rationalization of production methods - Ford liked to experiment - compelled the workers to either keep up the pace or quit. Many quit. "It cost us as I can remember about 38 dollars per man to break them in on the job," the employment manager at Ford's in 1913 recalled (Russell 1978:40). The turn-over approached 40% in that same year, which meant that the Company was spending over one and a half million dollars to train workers who might decide to quit after a week or less on the job. The jobs were supposed to be simple and easily learned, but the task was usually one that brought with it intolerable monotony and fatigue. To maintain a large, disciplined, and unskilled labor force, half of whom were immigrants, was one of Henry Ford's top priorities in the 1910s. Two managerial innovations were supposed to help solve the problem of labor turnover: the Ford Sociological Department and the Ford English School. The latter institution is usually described in connection with "Americanization", whereas the Sociological Department is reckoned as a means for administering the famous "Five Dollar A Day" wage that Ford announced in January 1914. I shall examine these two programs so as to reveal the close connections between them.

The high rate of labor turnover and the threat of union organization inspired Ford to take dramatic action. He released to the press on January 5, 1914 a "profit-sharing plan" that according to Ford, would "inaugurate the greatest revolution in the matter of rewards ever known to the industrial world". The official title for the program was "profit sharing", but it came to be known by its nickname, "the five dollar a day" - which was almost double the average wage for the factory worker at the time.

Ford's logic for the five dollar a day was not, as is commonly assumed, a mere economic incentive to the employee. For Ford the plan had a moral, even missionary, purpose. The program required an administration of the labor force in its leisure time as well as on the job. Ford appointed a social worker, John Lee, to head the Sociological Department, which was sometimes referred to as the Ford "welfare program". Its staff members visited the workers at their dwellings, over half of whom lived in rented rooms. There the "investigators" - they were not graduate students in sociology - observed the conditions and habits of each employee and conducted a questionnaire to determine if the person merited the benefits of working for Ford. Nearly 41,000 workers at the Highland Park factory received a visit from a member of the Ford Sociological Department in the period 1914-1916: surely one of the biggest research projects in the history of sociology, but one that never is recorded in the annals of the profession. In his monumental history of the Ford Motor Company, Alan Nevins described the method of the Sociological Department:

> Each investigator, equipped with a car, driver, and an interpreter, was assigned a district in Detroit, mapped to contain a due proportion of Ford workers, and if possible, a limited number of language groups. The subjects for inquiry made up a formidable list. Naturally, each worker was expected to furnish information about his marital status, the number of dependents and their ages, and his nationality, religion and (if alien) prospects of citizenship. Did he own his own home? If so, how large was the mortgage? If he rented a domicile, what did he pay? Was he in debt? And to whom? How much money had he saved, and where did he keep it? Did he carry life insurance, and at what premiums? His social outlook and mode of living also came under scrutiny. His

health? His doctor? His recreations? The investigator meanwhile looked sharply, if unobtrusively, so that he could report on 'habits', 'home condition', and 'neighborhood'. Before he left a given family, he knew whether its diet was adequate, whether it took in boarders - an evil practice which he was to discourage, and whether money was being sent abroad. All this information and more was placed on blue and white forms. The Sociological Department was nothing if not thorough (Nevins 1954 vol. I : 554).

Unfortunately those blue and white forms were lost in a fire. They would have provided abundant material for the study of Fordian sociology. The accumulated statistics of the Sociological Department's survey of the Ford workers have been preserved, according to nationality (See Tables, H. Ford Museum Archives).

The "Americans" in Ford's employ were those workers who had been born in the U.S.A. Whether their ancestors had arrived on the Mayflower or they were second-generation immigrants is not indicated. The category is simple and says nothing of ethnicity or regional background.

The Americans at Ford's were not distinguished for their good habits. The difference between the nationality groups with "good habits" and "fair habits" was about the same for all the nationalities. Few were seen to have "poor habits". Perhaps the Poles, the second largest group after the Americans needed a slight push towards moral improvement. Good habits included for example, having a savings account at a bank. Gambling was a poor habit, but apparently wide-spread among the workers.

Nationality itself did not appear to be a firm indicator of moral character, but it seemed increasingly important for Ford to pursue an Americanization program among his immigrant workers. The Sociology Department identified those employees who needed to learn English. Thus, "Language and Citizenship" was one of the significant aspects of the Department, which passed on the names and nationalities of prospective pupils in the Ford English School.

All of the Canadian and American workers spoke English - a total of 18,000 of the 41,000. Of the remaining 23,000 foreign-born, nearly 5,000 were found unable to speak English. The eastern European nationalities were clearly the most needy groups, averaging about one/third unable to speak English. Poles, Russians, and Serbians led among the Slavic non-English speaking, and Rumanians were perhaps the least trained in English. The Rumanians likewise had a relatively high proportion of workers with "fair habits" as distinguished from "good habits". Like the Poles, the Rumanians were in need of extra Americanization.

If saving at a bank indicated good habits, the Eastern European immigrants could be said to have excelled over the Americans. The Americans had well below the average of savings and the Rumanians, Russians, and Poles were above average. The high rate of savings, as many immigrants would acknowledge, was often meant to buy a return ticket to their homeland and bring back the fruits of one's labour in America. Or, the money could be sent to members of the family for passage to America. The "motives" for poor, fair, or good habits are not evident in the statistics, but one point ought to be clear: the notion of community, fundamental in the Chicago sociologists, is omitted from the Ford Sociology. Ford's sociological realism meant in effect that community had to be dissolved in the interests of industrial society. If "Homo Economicus" was Ford's ideal American, Ford recognized that such a man was not born, he had to be manufactured.

The sources of Fordian sociology do not lie in classical economic theory, surely not in university faculties. The most obvious source for the Sociological Department was the work by Frederick Taylor which we have already referred to: *The Principles of Scientific Management* (1911). It is likely that John Lee was familiar with Taylor's method. Taylor's experiments might well have served as the blue-print for the Ford Sociological Department.

When Taylor chose his candidates for research, he looked not only for those "physically able" to do the increased work, but those "whose character, habits, and ambition", seemed appropriate. Taylor's paradigmatic experiment dates from 1899 at the Bethlehem Steel Company. It does not have to do with modern technological invention but the shoveling, by hand, of pig-iron. The goal for the experiment was to increase the average load of pig-iron shoveled per man from 12.5 tons per day to 47 tons.

Frederick Taylor's own rendering of the experiment is best:

> Our first step was to find the proper workmen to begin with. We therefore carefully watched and studied these 75 men for three or four days, at the end of which we picked out four men

who appeared to be physically able to handle pig iron at the rate of 47 tons per day. A careful study was then made of each of these men. We looked up their history as far back as practicable and thorough inquiries were made as to character, habits, and ambition of each of them. Finally we selected one from the four as the most likely man to start with. He was a little Pennsylvania Dutchman who had been observed to trot back home for a mile or so after his work in the evening, about as fresh as he was when he came trotting down to work in the morning... This man we will call Schmidt (Taylor 1911, 1967).

The man called Schmidt, handpicked out of 75 laborers, actually *could* shovel 47 tons per day, and for Taylor, the experiment was a success. Henry Ford, however, was most interested in maintaining a labor force of tens of thousands of men whose names and habits differed from Schmidt. Taylor did stimulate Schmidt's capacity for shoveling by nearly doubling the conventional daily wage. There is every reason for thinking of Schmidt as a paradigm for the Ford factory worker. The Sociological Department, moreover, conducted on a vast scale the sort of inquiry into the worker's "character and habits" that Taylor managed with his four candidates. Thus, Ford magnified 10,000 times the Taylor experiment, demonstrating that the employer of labor could purchase not only production but reproduction.

To remake the immigrant workers into "Americans" was likewise a calling for Henry Ford in the 1910s. The Ford English School worked hand in hand with the Sociological Department. A manager of the English School described the methods and goals of the program:

> For their (the workers) intellectual improvement we have provided, among other things, the Ford English School. This is a school for foreigners in our employ, the enrollment averages about 2,000. The pupils are grouped in classes of about 25 to a class. The teachers are volunteers from the office and factory. There are over one hundred and sixty of them. Each class meets twice a week and the session lasts one hour and a half. Attendance is virtually compulsory. If a man declines to go to school, the advantages of training are carefully explained to him. If he still hesitates, he is laid off and given a chance for uninterrupted meditation and reconsideration. He seldom fails to change his mind.

There are over 50 nationalities in the factory and there may be as many nationalities in each class as there are men present, for we make no attempt to group them according to language and race. The fact is we prefer that classes be mixed as to race and country, for our one great aim is to impress these men that they are, or should be, Americans, and that former racial, national, and linguistic differences are to be forgotten (Ford Museum, Archives).

W.I. Thomas, whom we recall as one of the Chicago sociologists, was troubled that Americanization meant "the destruction of memories" (Gutman 1977: 75). With Henry Ford's sociology in Detroit, the destruction of memories was a goal, not a cause for worry.

The pedagogy of the Ford English School paralleled the method of the Sociological Department. The requirements of the Ford assembly line seemed to include a homogeneous work force where "racial", "national", and "religious" differences were eradicated. To impress the pupils of the importance of being remade into "Americans", the Ford English School designed a unique graduation ceremony. Again I quote from the text in the archives of the Ford Motor Company:

Not long ago this school graduated over 500 men. Commencement exercises were held in the largest hall in the city. On the stage was represented an immigrant ship. In front of it was a huge melting pot. Down the gangplank came the members of the class dressed in their national garb and carrying luggage such as they carried when they landed in this country. Down they poured into the Ford melting pot and disappeared. Then the teachers began to stir the contents of the pot with long ladles. Presently the pot began to boil over and out came the men dressed in their best American clothes and waving American flags (Ibid.).

The staging of Ford's immigrant workers undergoing a *rite de passage* has an almost surreal quality, akin to Charlie Chaplin's "Modern Times". But what has continued to impress me through the years as I have read and reread this precious document is that the times we are hearing about *were* "surreal." Modern managerial innovations were first clothed as scientific experiment and soon after as advertisement. Rationality in production became synonymous with Taylorism and Fordism... with ample reason.

The reality behind the scenes of Ford's melting pot and immigrant ship collided with the apparent harmony of mass Americanization. When one-thousand Ford workers, immigrants from Eastern Europe, took the day off to celebrate the Orthodox Church Christmas in January 1914, they returned to their jobs the next day to experience the real heat of the melting pot. The Detroit *Times* reported the event as follows:

When they returned for work on Thursday, they were escorted in one door of the plant, their numbers taken away, and they were then escorted out another door onto the street (Russell 1978).

How tempting it is to reduce the complexity of social history to the ethnographic *rite de passage*. On the stage, the immigrant enters the Ford melting pot, changes clothes, and comes out waving the flag. In reality, the unmelted worker has his number badge taken away and is fired. The point of the anecdote is almost too obvious: the melting pot was primarily a method for disciplining the labor force and preventing working-class communities from organizing. Americanization was a euphemism for discipline and homogenization. If, in the past, employers had housing built near the factory to keep convenient control over the work force, Henry Ford hit upon a new method: Send sociological investigators out to survey the workers in their residences. The sociological survey was a form of surveillance.

The project of the Ford Sociological Department and the Ford English School was at bottom a systematic rationalization of production - a far cry from the *verstehende Soziologie* of Simmel in Berlin or Park in Chicago. Ford's sociological method, moreover, had an appeal beyond the auto assembly line in Highland Park, Michigan. Even removing the melting-pot Americanization from the Ford project still left much to be accomplished in the name of modern rationalization. Besides marking the outbreak of World War I, they year 1914 also marked a first giant step towards the globalization of Fordism.[2] Fordism

---

[2] The spread of Fordism was marked for the first time on the European continent in Copenhagen, Denmark, where Henry Ford opened an assembly line in 1919. A large-scale production of Ford cars began in 1924 in the streamlined factory in Sydhavn. The Ford managers had to contend with trade-unions in Copenhagen.

and Taylor's "scientific management" won advocates in both the capitalist and anti-capitalist political economies following World War I. V.I. Lenin and A. Gramsci were notable Marxian proponents of Fordist rationalism. Lenin would adopt Fordism to facilitate Soviet industrial productivity, and for Gramsci Fordism expressed, not only technical efficiency but a radical transformation of the worker's moral character (Gramsci 1971).

Deliberating on the extent to which Fordian method was "rational," Gramsci thought he was able to separate the wheat (rationality) from the chaff (ideology). Gramsci discerned in Ford's practice an effort to create a disciplined work force via the inquiries into workers' habits and life styles. These inquiries were aimed at making a new sort of worker. They were not, as many claimed, an expression of Ford's "Puritanism." In Gramsci's words:

> It seems possible to reply that the Ford method is rational, that is, it should be generalized, but that a long process is needed for this, during which a change must take place in social conditions and in the way of life and habits of individuals. This cannot take place through coercion alone, but only through tempering compulsion, self-discipline with persuasion (Gramsci 1971: 312).

Conclusion: Beyond the Melting Pot?

In their study of ethnicity in New York City (1963) Professors Glazer and Moynihan assured their readers that ethnic membership and identity had survived the melting pot: "The point about the melting pot," they asserted, "is that it did not happen." Glazer and Moynihan were, as proper children of immigrants ought to be, aware of the ethnic bonds in their grandparents, their parents, and in their own childhoods.

Already in the 1910's the sociologists in Chicago, like the settlement-house workers before them, recognized the ethnic dimension among the immigrants. The Chicago sociologists often - but not always - viewed ethnicity as a hindrance to

---

One of the militant unions, the Saddlemakers', indicates the traditional, craft base of the union movement in Denmark. "Rationalization" of production in the long run has won the struggle over the craftsmen's "instinct" for autonomy. See Braverman 1974 and Veblen 1914.

acculturation in the New World. Tradition was a weight upon the immigrant, a kind of baggage that had to be abandoned in the interests of modern times.

Ethnic community, however, could also serve the members' struggles in the New World, and in its most perceptive moments, the Chicago School took the side of the immigrants in those struggles. Sociology *is Verstehen*. The Chicago School looked with wonderment at their city as an environment for growth, for exchange, for individuality.

Henry Ford's sociology was essentially anti-urban, just as it was radically industrial. Ford's was an extravagant asceticism, more advanced than any of the examples in Max Weber's portraits of "this-worldly asceticism" in the "capitalist spirit". As Gramsci had pointed out, Ford was not a Puritan. If anything, Ford was a latter day Victorian, which is not at all the same kettle of fish. For decades, Henry Ford had only one color for the cars that rolled off the assembly line: black. It was only after the threat of competition that he decided to diversify his colors. His taste in the popular arts, moreover, tended towards the homespun, small-town America, which Ford, more than any other American before Walt Disney, managed to promote. A thread of nostalgia runs through the projects of Henry Ford, the man who "put the World on Wheels," who "revolutionized" production and consumption.

The Henry Ford Museum in Dearborn (Greenfield Village) which I visited several times as a child, elevated small-town America's ingenuity and aesthetics. Though Ford glorified in the fact sometimes that so many of his employees were immigrants, I do not recall any objects in the Museum which honored that working population: no Polish folk costumes, no Italian ceramics. The melting pot did not reach the museum.

Ford, moreover, was not a generous patron of universities. He liked gymnasiums and swimming pools, and he funded the construction of the YMCA building in Highland Park.

The one possible anomaly in Ford's patronage is the fresco painting by Diego Rivera at the Detroit Institute of Arts, done in 1933 during the depth of the Depression. It was Henry Ford's son Edsel who actually arranged for the fresco painting, but one can puzzle, nevertheless, at Ford's *largesse*. What was Ford's motive for hiring the leftist Mexican to portray "Modern Times" with the Ford Rouge Plant at the very center? A later generation (of which I am a member) was taught to see the bitter irony and satire of

Rivera. But I now suspect that the artist and the patron both saw in modern technique and rationality a great leap forward. Fordism, then, exemplified modern managerial and productive capacity. Though fostered in an intensely moral zeal, Fordism *qua* rationality appeared as scientific neutrality, and it could appeal to both capitalist America and Soviet Russia. The essence of Modern Times for both camps was its rational efficiency, and that efficiency was to be the code for ethics and aesthetics.

Henry Ford developed that ethic on the assembly line, in his Sociological Department, and at the Ford English School. The power of that ethic was symbolized by the deceptive image of the melting pot, which some have said "did not happen." Not only did the melting pot *happen*, it *boiled over*. Though Ford's sociological inquiries of the 1910s have not found their way into the profession's history, they ought to be there on the sociological spectrum, most appropriately on the infra-red end of it. The Sociological Department at the Ford Motor Company ended its applied research in 1918. A new institution, the Ford Service Department, replaced the "social scientific" agency. "Ford Service" was another euphemism, because it was in fact a private police force that maintained discipline and hunted down potential trade-union organization. Detroit Sociology at the Ford Motor Company was therefore short-lived. Though there have been many social scientists who grew up in Detroit, there has never been a Detroit "School," as there was in Chicago.

Bibliography

Bates, M. (1953) "Ecology" in A. Kroeber, ed. *Anthropology Today*. New York.

Beynon, H. (1973) *Working for Ford*. London

Braverman, H. (1974) *Labor and Monopoly Capital: the Degradation of Work in the Twentieth Century*. New York

Cooley, C.W. (1909) *Social Organization: A Study of the Larger Mind*. New York.

Coser, L.A. (1971) *Masters of Sociological Thought*. New York.

Davis, M. (1980) "Why the U.S. Working Class is Different," *New Left Review*, 123.

Debouzy, M. (1972) "Le problème de l'immigration et l'histoire des États- Unis, *Annales: Économies, Sociétés, Civilizations*, 27.

Doctorow, E.L. (1972) *Ragtime*. New York.

Downs, L. (n.d.) *The Detroit Industry Frescoes by Diego Rivera*, Detroit Institute of Arts, Detroit.

Dreiser, T. (1900) *Sister Carrie*. New York.

Gerth, H.H. and Mills, C.W. (1958) *From Max Weber: Essays in Sociology*. New York.

Glazer, N. and Moynihan, P. (1963) *Beyond the Melting Pot: The Negroes, Puerto Ricans, Jews, Italians, and Irish of New York City*. Cambridge, Mass.

Gramsci, A. (1971) *Selections from the Prison Notebooks*, edited and translated by Q. Hoare and G.N. Smith. New York.

Gutman, H. (1977) *Work, Culture, and Society in Industrializing America*. New York.

Hannerz, U. (1980) *Exploring the City: Inquiries toward an Urban Anthropology*. New York.

Hartman, D.W. (1974) *Immigrants and Migrants: The Detroit Ethnic Experience*. Detroit.

Higham, J. (1963) (1975) *Strangers in the Land: Patterns of American Nativism, 1860-1925*. New York.

Knudsen, K. (1986) Kampen om samlebåndet: rationalisering og arbejder-bevægelse i Danmark i 1920'erne, *Den Jyske Historiker*, 35-36.

Kolko, G. (1976) Main Currents in Modern American History, chapter 3:*"The American Working Class: Immigrant Foundations."* New York.

Lipietz, A. (1982) "Towards Global Fordism?", New Left Review, 132.

Mommsen, W. (1974) Max Weber: Gesellschaft, Politik und Geschichte. Frankfurt am Main.

Nevins, A. (1954) Ford: The Times, the Man, the Company, 3 vols. New York.

Park, R.E. and Burgess, E.W. (1921) Introduction to the Science of Sociology. Chicago.

Park. R.E. and Miller, H.A. (1921) Old World Traits Transplanted, New York, reprint 1969. W.I. Thomas was the author of this work. Cf. Gutman 1977: 75.

Park, R.E. (1952) Human Communities: The City and Human Ecology. New York.

Powdermaker, II. (1968) "Field-work," International Encyclopedia of the Social Sciences, 5-6

Roberts, P. (1912) The New Immigration: a study of the industrial and social life of southeastern Europeans in America, New York.

Russell, J. (1978) "The Coming of the Line: the Ford Highland Park Plant, 1910-1914, Radical America, 12 nr. 3.

Schwartz, J.M. (1971) "Henry Ford's Melting Pot," in O. Feinstein, ed. Ethnic Groups in the City. Detroit.

Schwartz, J.M. (1985) Reluctant Hosts: Denmark's Reception of Guest Workers. Copenhagen.

Schwartz, J.M. (1985) "Moderne Tider: Hvorfor Immigranten Falder Mellem To Stole", Stofskifte, 12.

Simmel, G. (1950) The Sociology of Georg Simmel, trans. and ed. by K. Wolff. New York.

Simmel, G. (1959) Essays on Sociology, Philosophy, and Aesthetics, edited by K. Wolff. New York.

Taylor, F.W. (1911) (1967) The principles of Scientific Management. New York.

Thomas, W.I. and Znaniecki (1918) (1923) The Polish Peasant in Europe and America. New York.

Tönnies, F. (1957) Community and Society, trans. and ed. by C. Loomis. New York.

Veblen, T. (1904) The Theory of the Business Enterprise. New York.

Veblen, T. (1914) The Instinct of Workmanship. New York.

Vecoli, R.J. (1972) "European Americans: From Immigrants to Ethnics," International Migration Review, VI, Winter 1972.

Vidich, A.J. and Lyman, S.M. (1985) *American Sociology: Worldly Rejections of Religion and their Directions.* New Haven.

Weimer, D.R. (1966) *The City as Metaphor, "Heathen Catacombs: on Theodore Dreiser."* New York.

Wirth, L. (1928) *The Ghetto.* Chicago.

Wirth, L. (1938) "Urbanism as a Way of Life," *American Journal of Sociology,* XLIV, I.

Zorbaugh, H.W. (1929) *The Gold Coast and the Slum: A Sociological Study of Chicago's Near North Side.* Chicago.

Zunz, O. (1977) "Detroit en 1888: Espace et Ségrégation," *Annales: Économies, Sociétés, Civilisations,* 32, I.

Archival materials

Arbejderbevægelsens Bibliotek og Arkiv, Copenhagen.

Henry Ford Museum, Greenfield Village, Michigan.

I am grateful to my friend, Donald Gropman from Boston, Mass., who gave me some editorial help and encouragement in January 1987.

The Ford melting pot was an assembly line. Detail from Diego Rivera's fresco cycle in the Detroit Institute of Arts.

In Search of Homogeneity
Wilhelm Dilthey and the
Humanities

Peter Madsen

> Das philosophische Denken der Gegenwart dürstet und
> hungert nach dem Leben. Es will die Rückkehr zur Steigerung
> der Lebensfreude, zur Kunst usw. (Dilthey)

Dilthey is well known as one of the main contributors to the distinction between the natural sciences and the humanities understood as the social and historical branches of the academy. But
what interests me here is not so much his role in the general development of a hermeneutic approach as it is the way in which
he defined the framework of the humanities at a more specific
level. It may be too much to say that what we find in Dilthey is
some kind of paradigm that has been in power since, but I would
suggest for your consideration the argument that some of the
main features in Dilthey's late philosophy may help us to locate
weaknesses in the present concept of at least my own branch of
the humanities, i.e. literary criticism, even if the last two decades
or so have witnessed the emergence of something like a new
paradigm. However that may be, the interest of a glimpse of
Dilthey's thought is considerable from the historical point of
view of this conference.

Let me begin, then, with a quotation from Dilthey:

> Und die heutige Analyse der menschlichen Existenz erfüllt
> uns alle mit dem Gefühl der Gebrechlichkeit, der Macht des
> dunklen Triebes, des Leidens an den Dunkelheiten und den
> Illusionen, der Endlichkeit in allem, was Leben ist, auch wo
> die höchsten Gebilde des Gemeinschaftslebens aus ihm
> entstehen (Dilthey 1970a: 183).

This feeling that human matters are not marked by eternity may
have to do with the fact that Dilthey was approaching the end of

his own life, but his formulation - which is remarkably pessimistic in tone compared to the context in the book - has explicitly to do not only with personal life, but with the highest formations of community life. Among these highest formations are the national states. The heterogeneous dark desires may without distortion be associated with inner tensions of social character, like some kind of party in relation to the state (cf. ibid.: 211). "Gebrechlichkeit" is recognized as a current feature, even if Dilthey's main aim, and ideal, is harmony, i.e. to develop a kind of historical analysis, that underlines the unity of the cultural phenomenon in question, from the personality through the work of art to the national state.

What I'll try to do in what follows is to give a sketch of the way in which this search for harmony is inherent in the apparently philosophical or methodological structures of concepts - and the way in which this search for harmony may have repression as its practical counterpart. What I want you to keep in mind is the concurrent feeling of *Gebrechlichkeit*. I shall restrict my material to a few of Dilthey's latest writings, the last edition of *Das Erlebnis und die Dichtung* (1970b), and *Der Aufbau der geschichtlichen Welt in den Geisteswissenschaften* (1970a), both published in 1910, the year before his death.

1. The whole way through his historical introduction to *Das Erlebnis und die Dichtung* Dilthey is sketching the way in which poetry creates unity, even when the social reality is contradictory. The first small political-military communities expressed their common spirit in lyric poetry and "urwüchsiger Epik". The feudal societies were more complicated. *Their* common spirit was expressed in epic poetry too, poetry that was permeated by the "Geist" shared by social as well as political structures and expressed in religion. There was an opposition against the ideals of the church, but this opposition was "conditioned" by those ideals. This is the prevalent principle in Dilthey's historical sketch: contradictions are acknowledged but in various manners they are resolved by way of the cultural act, instead of heterogeneity culture creates homogeneity and harmony. Renaissance had as its proper genre the drama, i.e. Shakespeare. Here the contradictions are explicitly social, nevertheless:

> Aus dem Gegensatz, der die Gesellschaft dieser Zeit in eine aristokratische und eine niedere teilt, erhebt sich in Drama und Roman die Nebeneinanderordnung einer Welt vor-

nehmer Daseinsfreude und Lebensstärke und einer unteren, massiven, die nur durch den Humor dichterisch gestaltet werden kann (1970b: 9).

The result is that "aus dem Zusammenhang der Dinge erklingt eine unsichtbare Harmonie". And this is the "Geist der Zeit".

Then follows classicism with reason and rules, and thereafter *Aufklärung*. Here again the conflicts are explicit: between the ruling class and the bourgeoisie, between hierarchic coercion and freedom of belief, between despotism and political rights. But still *life* as such is the source of poetry, in this case the bourgeois sentimental drama and the comedy, the *Lustspiel*. This last one is Dilthey's favourite genre. Why? Because in the comedy the ambiguity of life is seen and enjoyed with amusement (*in Heiterkeit*). All the forces that were at hand for the literature of the *Aufklärung* were realized in the comedy - ambiguities and contradictions are dissolved under the aspect of amusement.

But then comes Goethe. The different versions of Dilthey's essay on Goethe has been analysed by Bernd Peschken in a very interesting book on which I am drawing in other respects (Peschken 1972), but in this context I am primarily interested in the last version of the essay. Here Goethe is regarded as some kind of main character, not only as far as poetry or literature is concerned but also as a human being and in relation to the humanities. The point is, that Goethe is so all-embracing, his integration of life-experience so comprehensive, that it corresponds to the task of the humanities, namely to *understand* the *objectivized experiences* of *life*. And here all the words are of a specific interest: understand, objectivize, experience, and life. But I'll come back to the more specific use of these words.

A few more crucial words may be introduced by the following quotation:

Und wenn nun die Erinnerung, die Lebenserfahrung und deren Gedankengehalt diesen Zusammenhang von *Leben*, *Wert* und *Bedeutsamkeit* in das *Typische* erheben, wenn das Geschehnis so zum Träger und Symbol eines Allgemeinen wird und Ziele und Güter zu Idealen, dann kommt auch in diesem allgemeinen Gehalt der Dichtung nicht ein Erkennen der Wirklichkeit, sondern die *lebendigste* Erfahrung vom Zusammenhang unserer Daseinsbezüge in dem *Sinn des Lebens* zum Ausdruck (ibid. 126, my italics).

Significance is given to things, when values meet life. That means that it is only through the specific human attitude to the natural surroundings that meaning is generated. This kind of meaning is mainly created through poetry, or it is poetry that articulates it at the highest level. Retrospective interpretation of the individual life is an activity, which is kindred with the poetic activity, but poetry reaches the general, the typical, and becomes therefore an expression of *ideals,* aims (*Zwecke*). The process of maturation is thus a still more embracing integration of *Erlebnisse,* i.e. a still more comprehensive integration of *Life.* But the capacity to embrace stems from the character of the soul of the poet and his level of phantasy.

> Es ist dann der Kunstgriff der grössten Dichter, das Geschehnis so hinzustellen, dass der Zusammenhang des Lebens selbst und sein Sinn aus ihm herausleuchtet. So erschliesst uns die Poesie das Verständnis des Lebens. Mit den Augen des grossen Dichters gewahren wir Wert und Zusammenhang der menschlichen Dinge (ibid.: 139).

This is no small task, but since it is not conceived as a critical criterion, but as a support for making Goethe the hero of poetry, it is mainly interesting in its general implications. And here it is important to stress Goethe's - according to Dilthey - universal character:

> Goethes (Universalität entstand) in dem *nacherlebenden Verstehen alles Menschlichen.* Wie alles Verstehen im Erleben gegründet ist, so ging es dann auch in ihm wieder zurück in die Erweiterung des eigenen Daseins (...). *Er verstand, indem er Fremdes zu seinem eigenen Leben in Verhältnis setzte, und das Verstandene wurde ein Moment seiner eigenen Entwicklung* (ibid. 167, my italics).

Goethe understood all human experience. All experience is the outcome of life, Goethe understood all alien experience and integrated it in his own work. So nothing is alien, all is embraced in Goethes work. And the reader who understands all of Goethe embraces all human experience. But Dilthey is even more explicit:

> Ebenso ist die Gesellschaft nach Ständen, Berufsarten, politischen Leistungen gegliedert und verbreitet sich in vielfachen Lebens-verhältnissen. *Überall sieht er Unveränderliches,*

Notwendiges. Grosse Bezüge, auf denen der Zusammenhang der geistigen Welt beruht, treten hervor (ibid. 175, my italics).

In other words: Goethe's phantasy transforms historical contradictions into necessary relations: "Seine Poesie versöhnt uns mit der Welt und verklärt sie" (ibid. 176-77). The question, or one of the questions, might be what extension this "we" is supposed to have. The French revolution threatened all the established, the result of this was the engagement in the totality. And this is Dilthey's main interest: the homogeneous totality. In the German tradition, Goethe's *Iphigenia im Tauris* holds the position as the sublime expression of this integration of the heterogeneous. The famous "lebt wohl" pronounced at the end by the barbarian king, when Iphigenia leaves his kingdom, is supposed to be the ideal expression of self-restraint generated by Iphigenia's humanity:

> In Tasso und Iphigenie schuf er sich dann eine andere ganz neue Form des Seelendramas (...) Seele wirkt hier auf Seele, und was draussen geschieht ist nur Gewand und Hülle (ibid.: 186).

2. One of the fundamental oppositions of categories in Dilthey's philosophy of the humanities is between *understanding* and *explanation*. But *understanding* has its foundation in experience, in *Erlebnis*. This is thus not only a fundamental, but a symptomatic term as well. Hans-Georg Gadamer had traced its history as word and as concept in *Wahrheit und Methode*, where he points out that the role of the term becomes still more important through the different versions of Dilthey's essay on Goethe. (The first version was from 1877). Even the title of the book is fairly late, from 1905. *Erlebnis* is already in the classic German idealism something else than reason, but it is at the same time something else than the *particular* sense-data, *Erlebnis* has to do with the totality. Gadamer finds even

> den Vorklang eines Protestes gegen die moderne Industriegesellschaft, der im Anfang unseres Jahrhunderts die Worte Erlebnis und Erleben zu Losungsworten von fast religiösem Klang aufsteigen liess (Gadamer 1965: 59).

As a concept *Erlebnis* has a strange middle-position in so far as it is related at the same time to the immediate and to the totality, it stems from life, but it is some *extraordinary* life-experience. The aesthetic experience is the prototype, since it is *above* ordinary life, but still is the expression of life. These are some of the main points made by Gadamer.

In English there is apparently no immediate difference - from the point of view of words - between *Erlebnis* and *Erfahrung*, but in the German tradition it is exactly in relation to *Erfahrung* that the specific features of the concept of *Erlebnis* become clear. *Erfahrung* has to do with reflexion and thereby with reason, *Erlebnis* is closer to intuition and the immediate. What is at stake here is one of the great oppositions in the intellectual traditions from the last two centuries. Dilthey's philosophy of the humanities is in clear opposition to the tradition of *Aufklärung* and its further developments in the 19th century - and this brings him close to other tendencies in our own period.

But let us first follow some of the steps of his reasoning in order to locate this problem in a more precise way. Behind the concept of *Erlebnis* lies obviously the concept of *Leben*. Dilthey wanted to get rid of preconceived conceptions, metaphysics and exterior restraints. The turn to this concept of *Leben* is an attempt to escape preconceptions, but it opens up for the introduction of irrational key concepts like depth and will to power.

In a way the natural sciences and the humanities have the same object, but under different aspects. Dilthey gives the example of the institutions of justice and the example of poetry (and the two spheres are not as remote as they may seem to be for a modern mind). The jail is a *physical* structure, but the physical fact that Mr. X is in this jail is a *cultural* fact, a question of *Geist*, too, since the physical structure of the jail is the result of certain political and administrative processes and the presence of Mr. X is the result of certain procedures formulated in laws and realized by some judges educated at certain universities. Poetry is blacking on white paper, a physical phenomenon that might be the object of study for the natural sciences, but for the humanities it is something else. Something parallel to the sentence to jail in as far as both are parts of a certain cultural situation, which is - in Dilthey's philosophy - fundamentally homogeneous. Somehow the formulations (*Äusserungen*) sum up into a common frame of mind or *Geist* of the epoch.

Life and life-experience are carried by so-called ideal unities, subjects of ideal character:

In der Familie, in verschiedenen Zwischenformen zwischen ihr und dem Staat und in diesem selber findet sich die höchste Ausbildung einheitlicher Zwecksetzung innerhalb einer Gemeinschaft" (1970a: 163).

But this leaves us with the problem: how is the relation between the many individuals and the ideal unity? Dilthey's solution might be named a kind of spiritual liberalism since the model at first sight is taken from something like the invisible hand of Adam Smith:

Ich finde das *Prinzip* für die *Auflösung* des *Widerstreites* in diesen Wissenschaften in dem Verständnis der geschichtlichen Welt als eines Wirkungszusammenhanges, der in sich selbst zentriert ist, indem jeder einzelne in ihm enthaltene Wirkungszusammenhang durch die Setzung von Werten und die Realisierung von Zwecken seinen Mittelpunkt in sich selber hat, alle aber strukturell zu einem Ganzen verbunden sind, in welchem aus der Bedeutsamkeit der einzelnen Teile der Sinn des Zusammenhanges der gesellschaftlich-geschichtlichen Welt entspringt: so dass ausschliesslich in diesem strukturellen Zusammenhang jedes Werturteil und jede Zwecksetzung, die in die Zukunft reicht, gegründet sein muss (ibid.: 167).

Several principles are in play in the attempt to create the picture of homogeneous culture. *First* the idea of a "common depth", which is completely irrational, *second* this *Geistes-Liberalismus*, some kind of inherent tendency towards unity, but *third* a more concrete homogenizing factor shows up: the state. And here another irrational feature from the period is in play, i.e. the will to power. The most explicit passage is this:

Die Klage über die Brutalität der Staatsmacht ist seltsam (...) Auf dem Boden dieses der politischen Organisation einwohnenden Machtwillens entstehen die Bedingungen, welche überhaupt erst die Kultursysteme möglich machen. So tritt hier nun eine zusammengesetzte Struktur auf. In dieser sind Machtverhältnisse und Beziehungen von Zwecksystemen zu einer höheren Einheit verbunden (Dilthey ibid.: 208-9).

And as an indication of wherefrom the dark forces, I referred to at the beginning, might stem Dilthey further on laments of the character of political organisations compared to states, named cultural systems:

> Wie verschieden ist die Struktur eines Kultursystems, in welchem ein gegliederter Leistungszusammenhang sich realisiert (...), von dem Wirkungszusammenhang in einer politischen Organisation (...) (ibid.: 211).

In the political organization we do not find an immanent law of development, the aims may change, the organization is compared to a machine that may be used to fulfil various tasks etc. Underlying is the historical fact, I suspect, that one of the most important political organizations at that time was the Social Democratic Party whose aims did not correspond too well to those of the German state.

But before I go into more detail in these matters I would like to return to the *Aufklärung*. In his book Dilthey gives an analysis of the way in which the German situation was the most happy realization of the aims of *Aufklärung*. In general his idea of the new epoch sounds like this:

> So entsteht die zusammengesetzte Struktur der staatlich organisierten Nation. Ihr entspricht eine neue innere Zentrierung dieses Ganzen. In ihm wird ein Wert für alle erlebt; das Wirken der Einzelnen hat an ihm ein gemeinsames Ziel. Die Einheit desselben objektiviert sich in Literatur, Sitten, Rechtsordnung und in den Organen des gemeinsamen Willens (ibid.: 214, my italics).

Among the factors involved in the happy German *Aufklärung* was a tendency to regulate life consciously through stable rules, and the seriousness of the Nordic peoples combined with a pondering need for composure (Besinnung), stemming from a turn towards the inward side of life etc. (ibid.: 220). In short: "Bindung und Pflicht" were parts of the German heritage from Luther. When the emancipation of the individual meets this tradition the result is an *obligation towards totality*: "Unbedingte Werte, Bindungen, Pflichten, Güter werden formuliert" (ibid.: 222).

But the construction of a cultural system is double-sided, it is not only a question of the *individual*, but also a question of the *state*:

> In dem Beamtentum, dem Militärwesen, den Finanzeinrichtungen liegt nunmehr das organisatorische Zentrum aller

Machtverhältnisse, und die Tätigkeit des Staates wird zu einer
treibenden Kraft der Kulturbewegung (ibid.: 222-3).

It is obvious that the central concern is the spirit of the epoch, but
that does not mean that Dilthey completely disregards the changing forces, the question of development from one epoch to another. But even if this question is formulated, it does not play any
considerable role in the argument. The dark forces of change are
considered as fundamentally heterogeneous elements in the
harmony between individuals and state. The complete harmony
between one individual and the state is demonstrated by a
character that shows up several times in the book, "ein Mann der
Tat" whose historical role is clearly supported by the man of con-
templation, i.e. Dilthey. I am talking of Bismarck. The role played
by Bismarck in Dilthey's book sheds new light on the remark
quoted above: "Die Klage über die Brutalität der Staatsmacht ist
seltsam etc." (ibid.: 208). Bismarck is taken as an example. One of
the main features behind his character is related to the land-
owning Prussian nobility, namely the sovereignty, the habit to
rule and lead, "Ungebrochenheit des Willens" (ibid.: 172). Another factor was the immense pressure laid on the political self-
esteem because of the situation of the state. All these factors take
part in the complex of influences (*Wirkungszusammenhang*), in
which the historian must situate Bismarck in order to understand how he became this great man of action and man of state.
This is briefly another example of the way in which Dilthey
wanted historical analysis to proceed. But it is more than that
since it is a description of the man, who was the leading character
in the creation of the German national state, in which Dilthey
found an example of the desired unity.

3. Let us now return to Dilthey's position in relation to the
history of philosophy: "Hegel konstruiert metaphysisch; wir
analysieren das Gegebene" (ibid.: 183). This is at the level of principles Dilthey's stand. But at the more concrete level it sounds
like this:

Er (i.e. Hegel) konstruierte die Gemeinschaften aus dem allgemeinen vernünftigen Willen. Wir müssen heute von der
Realität des Lebens ausgehen; im Leben ist die Totalität des
seelischen Zusammenhanges wirksam (ibid.).

Peter Madsen

The interesting point is not the interpretation of Hegel, but the fact that *Dilthey rejects the idea of a general will based on reason,* or to be explicit: the idea of a democratic society aiming at justice. At the same time he introduces, as I have mentioned, a concept of knowledge in the humanities that structurally corresponds to the poetic knowledge, it is intuitive and based on some inner correspondence:

> Leben, Lebenserfahrung und Geisteswissenschaften stehen so in einem beständigen inneren Zusammenhang und Wechselverkehr. *Nicht begriffliches Verfahren* bildet die Grundlage der Geisteswissenschaften, sondern *Innewerden eines psychischen Zustandes in seiner Ganzheit* und Wiederfinden desselben im *Nacherleben* (ibid.: 164, my italics).

This abdication of reason implies that Dilthey not only created a conceptual framework that separated the humanities from the natural sciences, but also brought the humanities close to the arts. An aristocratic theory of knowledge, it has been called.

As I said at the beginning of this lecture the idea of harmony was central in Dilthey's endeavours. If we bring together the points made until now, we get an imaginative structure marked by heavy repressions. The two main characters are *Goethe* and *Bismarck,* both represent harmonious integration, Goethe in his capacity to embrace all human experience, or at least all inner experience, and Bismarck in his capacity to create a state with the will to integrate, and if necessary repress opposition (as he did with the so-called socialist-law). Bismarck, too, is described as a master of understanding in his sphere:

> ein Mann der Tat wie Bismarck, (...) wird in der Kunst, hinter dem Ausdruck Absichten zu lesen, von keinem Ausleger politischer Akten und keinem Kritiker historischer Berichte erreicht werden (ibid.: 165).

Both heroes are masters in the field of the humanities, since they have supreme capacities in interpretation. And both correspond to the specific German tradition sketched by Dilthey: self-restraint and integration. Poetry and state represent ideals of integration. The university professor understands both and provides the philosophy of the humanities that corresponds to the historical situation, in as far as he exchanges *reason* for *life* and *freedom* for *state*. Erased are in this way the liberal tradition, the idea of a democratic society based on liberal rights and aiming at freedom and justice. In the name of *life* and as a declared fight against metaphysics Dilthey offers *eine gemeinsame Tiefe* and *die Brutalität der Staatsmacht* with a view of a more elevated unity of cultural systems.

Bibliography

Dilthey, Wilhelm (1970a) *Der Aufbau der geschichtlichen Welt in den Geisteswissenschaften*, Einleitung von Manfred Riedel, Suhrkamp Verlag, Frankfurt am Main.

Dilthey, Wilhelm (1970b) *Das Erlebnis und die Dichtung*, Vandenhoeck &Ruprecht, Göttingen.

Gadamer, Hans Georg (1965)*Wahrheit und Methode*, J.C.B. Mohr (Paul Siebeck), Tübingen.

Pescken, Bernd (1972) *Versuch einer germanistischen Ideologiekritik. Goethe, Lessing, Novalis, Tieck, Hölderlin, Heine in Wilhelm Diltheys und Julian Schmidts Vorstellungen*, J.B. Metzlersche Verlagsbuchhandlung, Stuttgart.

The Origins of Heinrich Wölfflin's Art History

Joan Hart

The process of the professionalization of the discipline of art history appears to have begun at the turn of the century in German-speaking countries - Germany, Austria, Switzerland - and spread to other nations, notably Great Britain and the United States.[1] The proliferation of museums began at that time, with the concomitant need for professional staffs to catalogue and maintain the collections, as well as educate the museum-going public. Universities began to recognize the need for this kind of professional at a time when enrollments in universities climbed and patronage of the arts became increasingly secular and bourgeois. Artists, too, were in transition, more often producing art for an educated, elite market, rather than the aristocratic and religious groups of the past. The scholarly apparatus for communication of new ideas about art - the art journal - also began to flourish at the turn of the century. Several scholars developed new constructs to deal with new ideas in art, the new status of the art historical discipline, and the growth of other new disciplines. August Schmarsow and Robert Vischer, who attempted a balance between aesthetics and the history of art, were prominent and now forgotten. The efforts of Heinrich Wölfflin to develop a new method for understanding art have not been forgotten. Nor have the efforts of the diverse members of the Vienna School of art history been lost. I will attempt to explain the reasons for Wölfflin's success in this paper.

The political situation in Germany at the turn of the century was a stimulus for the independence and recognition of art historians (Hart 1981; Paret 1981). Emperor William II was antipathetic toward modern art. Although he had engaged Hermann Grimm, Wölfflin's predecessor in the Chair of Art History at Berlin University, to educate his children about art, he took a

---

1 The concept of "professionalization" presented here is somewhat hypothetical, based on the author's own conclusions (Fawcett and Phillpot (eds.) 1976; Gombrich 1979; Kleinbauer 1971; Kultermann 1966; Paret 1981; Schlosser 1934).

very belligerent stand against modern art. In a famous incident in 1908, Emperor William II attended an opening of a modern art exhibition at the National Gallery of Art in Berlin, where he proceeded to insult Hugo von Tschudi, Director of the Museum and an avid collector of modern art for the Gallery, and Heinrich Wölfflin, head of the Art History program at the University - both Swiss. Von Tschudi resigned from his position in a glare of publicity, while Wölfflin basked in the notoriety of the incident. As a result, Wölfflin began to teach a course in contemporary German art. Modern art and modern art history tended to thrive in the face of official condemnation and disdain. 1908 was a year after the advent of Cubism, and about the time of Malevich's and Kandinsky's initial forays into the unknown territory of abstraction or non-objective art. Sezession movements of artists existed in Berlin, Vienna and Munich in the 1890s and generally thrived outside the official circles of art patronage. The artists do not appear revolutionary today, but, at the time, the art of Liebermann, Kollwitz, Feuerbach, even Böcklin, were too modern for a large part of Wilhelmine society. By 1908, the Berlin Sezession had exhibited the works of Nolde, Kandinsky, Feininger and Beckmann, but their work was then in transition, between the past and future. It was not really the style which society found objectionable, but rather the lack of nationalistic and chauvinistic subject matter. Furthermore, since some of the leading artists and gallery owners who exhibited their art were Jewish, the opposition against modern art in Berlin was particularly virulent in official circles. But neither art nor art history seemed to suffer very much. Wölfflin became so independent that in 1912 he was able to select as his successor in the Chair of Art History at Berlin University the first non-baptized Jew ever to hold a professorship at Berlin, Adolph Goldschmidt.

Perhaps it was not accidental that it was a Swiss who was innovative in the field of art history and who had the temerity to challenge the fear that modernism threatened the substance of society and its political structure, a belief widely held at the turn of the century, and still held in some countries. Who was Heinrich Wölfflin? He was no firebrand; in fact, he fits Fritz Ringer's depiction of the Mandarin professor like a glove (Ringer 1969). Wölfflin was born in 1864 in Wintertur, Switzerland to Eduard Wölfflin, a classical philologist, and Bertha Trolle Wölfflin, from a wealthy Basel family. Heinrich was the eldest of three children, none of whom married (Hart 1981).[1]

Over the course of his life, Wölfflin published four major books on art history, concerning the Renaissance and Baroque periods, in the North and the South. His first publication was *Renaissance and Baroque* (1888), his *Habilitationsschrift*, which was followed by *Classic Art* (1899), concerning art of the High Renaissance in Italy, *The Art of Albrecht Dürer* (1905), and *The Principles of Art History* (1915), for which he is best known. Wölfflin began teaching at Munich University as a *Privatdozent* in 1888, until 1893 when he was chosen by Jakob Burckhardt to be his successor at Basel University. In 1901, Wölfflin was invited to Berlin to succeed Hermann Grimm. Wölfflin stayed at Berlin University until 1912, when he moved to Munich to teach. He considered Munich to be a more free and easy cultural center, where his family still resided, where he had matriculated, and where his father had taught philology. Wölfflin, who was depressed by World War I and felt increasingly uneasy in Germany, left to return to his beloved Switzerland in 1924 where he taught at Zurich University until 1934, when he retired. He lived in Zurich until his death in 1945.

Wölfflin's teaching and writings had a profound effect on art historians and on the humanities in general (Hart 1982: 292-300). Even now, almost one hundred years after his first publication, it is not difficult to appreciate the impact of his new method of art history. Prior to Wölfflin's books, art history tended to be without method, often anecdotal; authors listed artists and described their work, and were sometimes guided by ill-defined notions of style and period. Wölfflin forever changed this inattention to method. His contribution to methodology appears very simple, and it is perhaps because of its simplicity that it had and continues to have such an impact. In *Renaissance and Baroque* (1888), he proposed to compare the two period styles to determine the broad characteristics which defined them. In order to determine these characteristics, Wölfflin initiated a method of formal analysis, analyzing the formal structure of Renaissance and Baroque works of art to determine how they differed. For example, in this early work, he distinguished the differences in Renaissance and Baroque ar-

---

[1] My information on Wölfflin derived primarily from my research at the *Wölfflin Nachlass* in the *Handschriftenabteilung*, Basel University Library, Basel. Following American academic practices, almost all the German has been translated for this article.

chitecture with these terms: painterliness, grandness, massiveness and movement. The Baroque style had these features while the Renaissance style did not. This type of comparative stylistic analysis was the foundation of his method. He used it to compare Quattrocento and Cinquecento art in *Classic Art*; and to describe Dürer's artistic development from a purely Northern Gothic style gradually adapted to incorporate Italian classic elements. Finally, in *The Principles of Art History*, of 1915, Wölfflin used a construct of five opposed formal characteristics to describe the difference between Renaissance and Baroque art: Renaissance art was linear, closed, planar, clear and had a multiplicity of foci, while Baroque art was characterized by painterliness, openness, deep spatial elements, relative unclearness, and a unified focus. The persuasiveness of this method derived primarily from Wölfflin's genius in describing works of art with sensitivity and precision, despite the schematization of the styles. Furthermore, he used an empirical and inductive approach, accumulating examples so that, in the end, it was difficult to deny the correctness of his results.

Underlying the method was an explanatory substructure which was intended to ground his results. He began, in *Renaissance and Baroque*, with an explanation for the change in style from Renaissance to Baroque: a change in the psychology or "mood" of a people. In the early 1890s, Wölfflin ascribed stylistic changes to a fundamental shift in perception and cognition, discussing this idea in *Classic Art* and *The Principles of Art History*.

The method and the explanation for change in artistic styles seem very simple, not requiring much explanation to account for their occurrence in Wölfflin's work. However, they were quite revolutionary in their time and they continue to exert a profound influence not only in the history of art, but in other disciplines, also. The psychologism of Wölfflin's theory still has strong adherents, such as E.H. Gombrich and Rudolf Arnheim in art history, and they in turn have generated much discussion in other fields, notably literary history.

Because the method and theory of explanation appear in Wölfflin's first book, *Renaissance and Baroque* of 1888, the sources of both must have been acquired in his university days. By the time Wölfflin was ready to attend the university in 1882, his father was a well-known Professor of Classical Philology at Munich University. Heinrich was a precocious student: first in his class at the Gymnasium, he had mastered six languages, and was deeply interested in history, literature, philosophy and art.[1] Like most students, he moved from one university to another, seeking out the professors with whom he wanted to study. He attended Basel, Munich and Berlin Universities between 1882 and 1886. He began his university education at Basel. He had several reasons for selecting Basel but the primary one was to study cultural history with the famous Jakob Burckhardt. Throughout his university training, Wölfflin's goal was to become an historian of culture. From the outset he believed that philosophy would provide a sound basis for cultural history and he specialized in philosophy, taking his degree in philosophy at Munich University in 1886, with his father presiding at the ceremony. In 1884, Wölfflin wrote to his parents:

> *Philosophy* is the highest constant for me. It unites an entire age. I would choose it as the foundation for every higher cultural history: philosophy and history complement each other reciprocally. The object of both is man, the whole thinking, feeling mankind; the former analyzes it, the latter gives it history.[2]

Later Wölfflin was accused of forgetting the human element in his categorization of art, but he never really lost sight of his goal.

Burckhardt's humanistic influence is manifest in this quotation. Wölfflin studied Medieval and Renaissance art and history with Burckhardt, but increasingly found his association with Burckhardt to be unsatisfactory. From the beginning Wölfflin sought a method for the study of cultural history which Burckhardt did not seem to offer.[3] Burckhardt assigned books in his

---

[1] Letter from Eduard Wölfflin to Burckhardt, from Munich, December 7, 1882 (Gantner ed. 1948).

[2] *Wölfflin Nachlass*, III A 69, Letter from Wölfflin to his parents, from Basel, May 11, 1884. "Am höchsten aber steht mir die *Philosophie*. Sie fasst je eine ganze Zeit zusammen. Ich würde sie entschieden als Grundlage für jede höhere Kulturgeschichte wählen; Philosophie und Kulturgeschichte ergänzen sich gegenseitig, beide haben zum Object den Menschen, den ganzen denkenden, wollenden, fühlenden Menschen. Die eine analysiert ihn, die andere giebt seine Geschichte."

[3] The theme of finding a method is pervasive in Wölfflin's notebooks and letters during his university years, 1882-1886. In 1885, Wölfflin wrote his father from Berlin: "There are two methods, the philosophical and the philological and both are right." (*Wölfflin Nachlass* III A 117). Wölfflin's notes from Burckhardt's course still exist: "Geschichte der neuern Zeit, 1450-1598," from Winter Semester 1882-1883 (*Nachlass* II A 6).

courses that did offer new approaches; he apparently admired Henry Thomas Buckle's *History of Civilization in England* which appeared between 1857 and 1861.[1] Buckle applied positivism to English history with a vengeance. Wölfflin briefly expressed great enthusiasm for Buckle, but even more for August Böckh, a proponent of the inductive method in the humanities, who had taught philology and rhetoric at Berlin University for fifty years.[2] Böckh subscribed to an inductive approach and a hermeneutic method for the study of philology in his posthumous book, *Encyklopädie und Methodologie in den philologischen Wissenschaften*. Böckh was extremely influential, having taught many of the great minds of the nineteenth century, including Burckhardt, Nietzsche, Karl Otfried Müller and Wilhelm Dilthey, all of whom had an influence on Wölfflin. Furthermore, philology was the most important discipline in the German academic system in the late nineteenth century (Ringer 1977: 332-347). Philology then did not have the same limitations that it does now. Böckh viewed philology as embracing world history, since it provided a basis for scrutinizing historical and literary texts, and the method of textual analysis was hermeneutics. Böckh believed that philology was independent of but complementary to philosophy, both in method and subject.

Philology had special importance for Wölfflin, not only because it was the basis of much of his education, but because his father was a leading figure in the discipline, an originator of the *Thesaurus Linguae Latinae*.[3] The method of hermeneutics was of continuing interest to philologists throughout the nineteenth century and surfaces in Wölfflin's work in an interesting way. His history of art was also influenced by philological methods other than hermeneutics. Eduard Wölfflin was a comparative philologist and it appears to have been at his suggestion that Heinrich began to use the comparative method that became the hallmark of his art history.[4]

---

[1] *Wölfflin Nachlass*, Notebook 6, 1883, p. 49, Buckle was among the few books assigned by Burckhardt. Wölfflin's manifesto of study to his father referred to Buckle: December 1, 1882, from Basel (*Nachlass* III A 46.)

[2] Burckhardt assigned Böckh's book (Böckh 1886).

[3] "He (Eduard Wölfflin) took a great interest in methodological exactitude" (Wölfflin 1933: 330). *Nachlass* IV 1866, Letter from Eduard Wölfflin to Heinrich, from Munich, December 10, 1882.

[4] *Nachlass* Notebook 9, p. 46. Wölfflin wrote an outline for a comparative history of art based on A. H. Sayce's book, *The Principles of Comparative Philology* (Sayce 1874).

During Wölfflin's first two years in the Universities of Basel and Munich, he was primarily influenced by scientific thinking - that of Buckle, Böckh and his father - in his search for a method for the study of cultural history. Wölfflin inquired:

Can history, which was until now only an *experientia*, be raised to a science in which, following the model of the natural sciences, one extracts from the profusion of facts the great laws of spiritual development in the human race?[1]

This question remained in Wölfflin's mind throughout his life, but he found a counter-tradition to it in his study of philosophy. The enthusiasm for positivism had never been strong in Germany, and Wölfflin, in the course of his studies at Basel, Munich and Berlin Universities, came into contact with professors and authors who remained unconvinced that the methods of the natural sciences were appropriate for any of the cultural sciences or *Geisteswissenschaften*. By the 1880s, the reaction against positivism was in full swing, particularly among philosophers (Ringer 1969: Chapter V).

Aesthetics was one of Wölfflin's strong interests. Wölfflin studied aesthetics with Johannes Volkelt at Basel University in 1884. Volkelt's assignment to Wölfflin was to consider the nature of form in art.[2] This was to become a lifelong preoccupation. Volkelt's most important work in aesthetics was a small treatise entitled *Der Symbolbegriff in der neuesten Aesthetik* (1876), in which he traced the concept of symbolization chronologically, in the works of Robert Zimmermann, Hegel, Herman Lotze, Friedrich Theodor Vischer and his son Robert Vischer, Gustav Fechner and others. To Volkelt, Lotze and the Vischers, "symbolization" meant a relationship between the viewer and a work of art which was empathetic; the subject imbued the object with life. They explained this symbolic and empathetic relationship by means of psychological and physiological processes. In his dissertation of 1886, *Prolegomena zu einer Psychologie der Architektur*, for the Philosophy Faculty at Munich University,

---

[1] *Nachlass* III A 46, letter from Wölfflin to parents, from Basel, December 1, 1882: "Kann die Geschichte, die bisher nur eine *experientia* war, zur Wissenschaft erhoben werden, indem man, nach dem Vorgange der Naturwissenschaften, aus der Fülle der Tatsachen die grossen Gesetze der geistigen Entwicklung des Menschengeschlechtes herauszieht?"
[2] *Nachlass* III A 71, III A 72, letters from Wölfflin to his parents, from Basel, May and June 1884, referred to his assignment from Volkelt.

Wölfflin attempted to make concrete this anthropomorphizing of the work of art.[1] He suggested ways in which the human body was symbolized in architecture, by comparing features of both. Unlike the philosophers of symbolization, Wölfflin cited the experimental results of psychologists Fechner and Wundt to support his theory. He had become interested in their psychological research when studying with the philosopher Wilhelm Dilthey at Berlin University in 1885 and 1886.[2] Psychology was still considered a part of philosophy and Wölfflin studied it with three professors: Volkelt who was to become one of the originators of *Einfühlungspsychologie*, Ebbinghaus at Berlin whose pioneering research on memory is still quoted, and Friedrich Paulsen who was a neo-Kantian philosopher at Berlin. Thus Wölfflin was familiar with all the trends in the new discipline.

The application of psychological concepts in philosophy began among neo-Kantian philosophers (Ermarth 1978: 73-75; Willey 1978: 74, *passim*). The trend among German academic philosophers to return to Kant for the foundation of a critical philosophy began in the 1850s, reached its height in the 1890s and disappeared after 1914. During this period, Hegel was viewed as ultra-conservative, the voice of reaction and the status-quo. Kant, by contrast, had offered his philosophy as a starting point, which led later philosophers to believe that "to understand Kant is to go beyond him." Some nineteenth century philosophers believed that the introduction of psychological concepts into philosophy represented an extension of Kant's categories. Herman Lotze, Eduard Zeller and Friedrich Theodor Vischer were the early proponents of this idea: they identified Kant's *a priori* forms with the operations of brain physiology and called this new formulation "psychologism."

Wölfflin's most important mentor in philosophy was Dilthey, who was influenced by neo-Kantianism. Wölfflin, still seeking a method for the study of cultural history, found it in the work of Dilthey. Evidence of Dilthey's influence on Wölfflin is found throughout his notebooks and letters to his family.[3] Dilthey was

---

[1] Wölfflin's dissertation is reprinted in Heinrich Wölfflin, *Kleine Schriften 1886-1933* (Wölfflin 1946: 13-47).

[2] *Nachlass* II A 9, Wölfflin's notes from Wilhelm Dilthey's course "Logik und Erkenntnistheorie," winter semester 1885-1886, University of Berlin.

[3] *Nachlass* III A 75-107, Letters from Wölfflin to his parents in 1885; *Nachlass* Notebook 9, 1885, p. 9; Wölfflin stated that Dilthey's book is "a foundation for the study of society and history." (Dilthey 1883; Ermarth 1978: 73-75).

later instrumental in helping Wölfflin obtain the offer to come to Berlin in 1901. When Wölfflin began his studies with Dilthey, the latter had recently published his first major philosophical treatise, *Einleitung in die Geisteswissenschaften* of 1883, in which he contrasted positivism, the method of the natural sciences, with his alternative method for the cultural sciences. Dilthey formulated a coherent response to the positivists - Comte, Buckle and Mill - who had proposed applying the methods of the natural sciences to the disciplines focussing on man and his productions.

The main part of Dilthey's book consisted of an examination of the ascendancy and decline of metaphysics throughout history. He proposed substituting a new epistemology for the dead metaphysics and the Positivism that had taken its place: an epistemology of the cultural sciences based on a descriptive psychology. Dilthey contended that the one central element that distinguished the cultural sciences from the natural sciences was "consciousness." His conception of consciousness was historical; through it one could reconstruct and re-experience another person's inner world. Dilthey thought that consciousness was the basis of all human knowledge and that only by investigating consciousness would an explanation of knowledge be possible. The method he proposed was antithetical to the naturalistic, causal models of positivism, which he felt could not be used to understand the inner life and experience of man.

Mediating between consciousness and man's understanding of it was logic, according to Dilthey. He believed logic could be an interpretive device, providing a structure for lived experience. The logic Dilthey had in mind was not rigorous, mathematical, or *a priori* , but rather a logic dependent on language and its content, changing throughout history. Thus, in 1883, Dilthey proposed that the foundational science of all the cultural sciences would be a descriptive psychology and logic would provide the structure for the experience of consciousness.

The parallels between Wölfflin's ideas and Dilthey's are clear: like Dilthey, Wölfflin (beginning with his dissertation of 1886) believed that psychology would provide the regulative laws, the foundation, for all the humanities. They believed in providing a logical structure for their theories; they avoided causal statements; they were antipathetic to materialism and Hegel's philosophy; both were concerned with bringing philosophical idealism into harmony with nineteenth century science, and they believed in the centrality of consciousness.

Joan Hart

The influence of Dilthey on Wölfflin was as evident in Wölff-lin's dissertation, as the already noted influence of Volkelt and the psychologists, Fechner and Wundt. The purpose of exploring Wölfflin's earliest writing is to demonstrate the origins of his thought which became more convincing in later publications, while retaining the same sources. The concept of "symbolization" seems particularly bizarre today, as used by aestheticians, but it may also have had an effect on the development of the later iconographical theories of Aby Warburg and Erwin Panofsky, and Freud's psychoanalytic theory.

Wölfflin constructed his dissertation, *Prolegomena zu einer Psychologie der Architektur*, on a framework of neo-Kantianism. It included an idiosyncratic interpretation of Kant's *Critique of Pure Reason* (Wölfflin 1946: 29). Wölfflin believed that Kant defined a system as evolving organically and harmoniously, from within, like animals develop; the "system" was a unity of parts, coordinated by an idea expressing the goal and form of the whole. The regulating principle was the organic analogy; that is, man comprehended other objects in terms of his own physical organization. This kind of argument, which became more preva-lent after 1900 and became known as *Einfühlungspsychologie*, was an attempt to solve the perceived discontinuity between ex-perience and the object of experience. The analogy provided a link without requiring a causal relationship. This theory of em-pathy can be regarded as another aspect of the trend to counter positivism at the turn of the century. Wölfflin invoked Kant as the source of the idea.

The two main ideas in Wölfflin's dissertation were: first, to give a more modern and scientific basis to the traditional theory that architecture is organized in accord with the proportions of the human body, and, second, to provide experimental psycho-logical evidence to demonstrate that the forms of buildings di-rectly communicated organic, human laws. His main thesis was that architecture was expressive to us, because it expressed in the same way as man. Buildings were analogous to man in both structure and function. Using the physiognomic analogy, Wölff-lin discovered the impulse to engender life in the forms of the facades of buildings: windows were like eyes, a cornice was like brows, and so on. Architectural terminology supported the analogy. The evidence that seemed to prove the thesis came from the experiments of Wundt and Fechner. For example, Wundt demonstrated that subjects associated lines and colors as if they expressed the same thing: the hasty back and forth movement of a zigzag was associated with red, while blue was associated with curves. Fechner performed experiments that indicated that slim subjects preferred slim proportions. This likeness of feeling could be easily communicated with architectural forms. It was a short step from this evidence to the conclusion that whole nations or groups expressed their character and the mood of the time through their architecture - and not only in architecture, for the same human impulse was first expressed in the form of shoes, costume, decoration and only later in the fine arts. Here we have an argument supporting the effect of mass culture on elite culture.

The great hopes for psychology were reinforced by experiments which seemed to prove the prevalent aesthetic notions of the day. But still more was anticipated. Wölfflin believed (in 1886) that psychology would eventually prove that "the organization of the human body is the constant denominator of all change." Although Wölfflin retained his belief in psychology, the form of his belief changed, just as the new discipline of psychology changed. Wölfflin obviously could not continue to believe that the constancy of the human organization was the foundation of artistic expression, because art changed while man's organization did not.

Like Dilthey, Wölfflin increasingly inclined toward an empirical examination of the evidence. A statement by Dilthey could be a statement of their common goal[1]:

The task of our generation is clearly before us: following Kant's critical path, but in cooperation with researchers in other areas, we must find an empirical science of the human mind. It is necessary to know the laws which rule social, intellectual and moral phenomena. This knowledge of laws is the source of all the power of man, even where mental phenomena are concerned.

---

[1] From "Die Dichterische und Philosophische Bewegung in Deutschland 1770-1800" (*Antrittsvorlesung* in Basel, 1867): "Die unsrige ist uns klar vorgezeichnet: Kants kritischen Weg zu verfolgen, eine Erfahrungswissenschaft des menschlichen Geistes im Zusammenwirken mit den Forschern anderer Gebiete zu begründen; es gilt, die Gesetze, welche die gesellschaftlichen, intellektuellen, moralischen Erscheinungen beherrschen, zu erkennen. Diese Erkenntnis der Gesetze ist die Quelle aller Macht des Menschen auch gegenüber den geistigen Erscheinungen." (Dilthey, *Gesammelte Schriften*, V, 27, quoted in English by Ermarth 1978: 142).

Joan Hart

One of the problems facing late nineteenth century philosophers was the resolution of the conflict between positivism and idealism. It was thought that the empirical study of the mind might provide a solution, as well as providing a foundational science for all *Geisteswissenschaften*. Reading Eduard Zeller and Friedrich Albert Lange, as well as studying with Friedrich Paulsen, could only have confirmed for Wölfflin the neo-Kantian position that psychology was the proper basis of Kant's *Critique*.[1] Lange's *History of Materialism* was a critique of materialism in which he discussed the epistemological limits of science while, like Zeller, proposing that the discoveries of physiologists were an advance on Kant's theory. Friedrich Paulsen, who also taught philosophy and psychology to Wölfflin at Berlin University, was a biographer of Kant. He was convinced that positivism was unable to account for the meaning and value of human experience. The specific features of Wölfflin's psychologism derived from Dilthey - both believed psychology was the foundation of the historical sciences, but that psychology could not be compared to the foundation of the natural sciences, mechanics.[2] Mechanics was ahistorical, while psychology was historical. This was a crucial distinction for both.

Wölfflin's next publication, *Renaissance and Baroque* (1888), owed even more to Dilthey. In a notebook entry for the preparation of this book, Wölfflin wrote about Dilthey's philosophy[3]:

---

[1] Wölfflin discussed Zeller and Lange in *Nachlass* Notebook 8, pp. 41, 50, 104, and Notebook 9, 1885, pp. 32, 118, *passim*. Willey, *Back to Kant*, discussed Zeller, Lange and Paulsen. Wölfflin, also, read neo-Kantians who were opposed to psychologism, Otto Liebman (*Kant und die Epigonen*) and Wilhelm Windelband. *Nachlass* Notebooks 12 and 13, 1885-1886, were devoted almost exclusively to Kant and neo-Kantian interpretations of Kant's philosophy.

[2] *Nachlass* Notebook 14, 1886, p. 158. Wölfflin may have been thinking of Rudolf Hermann Lotze, who believed that the mind functioned mechanically (*Mikrokosmos*). Wölfflin compared mechanics to psychology as foundational disciplines, but recognized the profound differences between them, also. (Wölfflin 1946: 45).

[3] *Nachlass* Notebook 14, 1887, p. 160: "Dilthey: Mein Standpunkt ist der einer Erfahrungsphilosophie, welche auch die Tatsachen der innern Erfahrung unbefangen zu gewahren und den Ergebnissen des Studiums der Aussenwelt gegenüber zu schützen strebt. Ich erkläre daher im Gegensatz gegen Hegel die Entwicklung der Philosophie nicht aus den Beziehungen der Begriffe aufeinander im abstrakten Denken, sondern aus den Veränderungen im ganzen Menschen nach seiner vollen Lebendigkeit und Wirklichkeit. Gilt auch für Kunst."

(his) viewpoint is based in a philosophy of experience which also preserves impartially the facts of inner experience and strives to protect the opposite, the products of study of the exterior world. In opposition to Hegel, therefore, I explain the development of philosophy not out of the relationships of concepts in abstract thought, but out of changes in all humanity due to man's whole animation and reality. Also valid for art.

In his notes, Wölfflin then proposed a way of creating a universal interpretation of art: by means, first, of a history of vision, second a history of the "feeling for form" (*Formgefühl*), and, third, a history of feeling and taste. These correspond to the three classes of statements in Dilthey's *Einleitung:* descriptive, explanatory and normative (Dilthey 1883: 26). Dilthey stated that united they defined his concept of *Verstehen* or understanding.

One could continue to trace the influence of Dilthey in Wölfflin's thought. The main features of Wölfflin's history of art remained very similar with each new publication: Burkhardt's love of the Renaissance is evident in all of them, Eduard Wölfflin's comparative method is found in all, the rejection of positivism and continuing psychologism are retained. In the 1890s, the form of the psychologism altered in Wölfflin's writings. Adolf von Hildebrand, the neo-classic sculptor from Munich, met Wölfflin in 1889 in Florence and read Wölfflin's dissertation.[1] In 1893, Hildebrand published a small volume called *The Problem of Form in the Fine Arts* which had a surprising impact on a number of art historians, Wölfflin and Alois Riegl among them.[2] Hildebrand's main contribution was a direct connection between perceptual behaviour and artistic creation. He justified and described his favourite art form, the classical relief, in terms of a *Fernbild* or distant image. When one views a far object, the eyes almost see parallel, no longer focus, and objects seem flatter, two-dimensional and unified. From a near viewpoint, the eyes must fix on one point at a time, roam, change accommodation and focus, and this was impossible to recreate in art. Hildebrand distinguished between a visual "effect" and an objective form, in this way recognizing that vision does not always correspond to an

---

[1] For a discussion of their first meeting (Hart 1981: 234-36). *Nachlass* III A 226, Letter from Wölfflin to his family, from Florence, June 14, 1889, recorded this meeting and is translated in Hart's dissertation.

[2] Hildebrand's influence was discussed in detail (Hart 1984: Band 1: Sektion 1).

objective form, for example when lines of equal length are placed so they appear to be unequal.

Hildebrand's book had an enormous influence, because he convincingly demonstrated how vision is manipulated by artists and the perceptual principles that led to certain effects. Wölfflin already believed by 1890 that vision was at the root of artistic change. Hildebrand gave a scientific basis to his theory. The general psychology of Dilthey was supplemented by the idea of an independent, autonomous artistic development dependent on perception with its own inner laws. In Wölfflin's book, *Classic Art*, he gave three reasons for the change in style from early to high Renaissance art: first, patronage changed; second, a new concept of beauty evolved; and, finally, perception changed. Wölfflin meant "perception" both in a narrow sense and in a general one, for he firmly believed that different generations actually perceived differently, that is, saw with different organs. Change in perception became the reason for change in art, supplanting his earlier notion of the human organization. This endogenous cause of change in artistic styles was prior to exogenous societal changes.

Dilthey gradually discovered the contradiction in positing psychology as the foundational discipline of the cultural sciences, but Wölfflin did not. If perception itself has its history, how can it provide a foundation for the history of art? Once it is transformed into history, perception is unable to ground history. In Wölfflin's theory, perception and psychology have entered the hermeneutic circle. To quote Wölfflin[1]:

> Since the entire history of perception (history of ideas) must transcend mere art, it is evident also that such national differences of the 'eye' are more than a matter of taste: determining and determined, they contain the bases of the whole world-image of a people (Wölfflin 1929: 256).

---

[1] Erwin Panofsky heard or read Wölfflin's Inaugural Address to the Prussian Academy in 1911, which was a summary of ideas that were the basis of the *Principles of Art History* of 1915. Panofsky published an article, "Das Problem des Stils in der bildenden Kunst", reprinted in *Aufsätze zu Grundlagen der Kunstwissenschaft*, Berlin, 1964, first published in 1915. Panofsky believed Wölfflin was using a nativist conception of perception which neglected content and expression in works of art (*seelisch*). However, Wölfflin did not intend such a restricted concept of perception and replaced the term *Sehgeschichte* to *Vorstellungsgeschichte* in *Kunstgeschichtliche Grundbegriffe*.

In Wölfflin's and Dilthey's thought, the idea of a foundational science became an unquestioned assumption (Ermarth 1978: 233-234). Dilthey moved on from psychology to hermeneutics as the basis of the cultural sciences. In some ways, Wölfflin, probably without being aware of it, again echoed his old mentor. His education in philology was thorough, due to his father's influence and beacuse it was so pervasive in the educational system. Throughout the nineteenth century, hermeneutics was a primary means of interpreting texts, and gradually came to be seen as a means of understanding experience.

A hermeneutic theory has a few unchanging components: It has been called "the art of understanding," a mental process of re-experiencing and reconstructing the past.[1] Understanding is dialectical, or comparative, since it can be attained only through the interaction between past and present viewpoints. Understanding is circular: the process of comprehending the part, then the whole, and *vice versa* is discontinuous, not logical, but sequential and open-ended. It is contextual and historical, allowing for the subjectivity of the interpreter by stressing the reconstructive process of understanding. Objectivity - total objectivity - is not a goal of hermeneutics.

In Wölfflin's lifetime, hermeneutics developed from an inductive method of criticism and interpretation in August Böckh's teachings, to an epistemological basis for the *Geisteswissenschaften* in Dilthey's thought, to a phenomenology of Being in *Sein und Zeit*, in which Heidegger transformed the epistemological problem into an ontological one.

Wölfflin stated the circularity of his thought without apology: on the one hand, preconceptions from past interpretations always condition what we see and understand, and, on the other hand, present perception limits how we interpret the past. Seeing is always "seeing as," context-bound. He could never decide whether the art, as "forms of perception," determined perception and the worldview or whether the latter conditioned the art. At the same time Wölfflin's method in all his books was empirical: He always began by marshalling the facts, gathering them together in a systematic way, and ended with inductively derived hypotheses. But the older Wölfflin became, the less positive he was that psychology would provide laws of human and artistic development. Increasingly, the hermeneutic statements Wölfflin

---

[1] The literature on hermeneutics is, by now, very large (Couzens Hoy 1978; Gadamer 1975; Habermas 1971 and 1979; Palmer 1969; Rorty 1979).

made to explain the phenomena he derived by empirical means were not consistent with his method:

> No one will want to assert that 'the eye' accomplishes developments in itself. Determining and determined, it always overlaps into different spiritual spheres. There is no optical schema that, only resulting from its own premises, could be imposed on the world like a dead pattern. But if one also at all times sees as one wants to see, that does not exclude the possibility that a law exists throughout all change. To know this law would be the main problem, the basic problem of a scientific art history (Wölfflin 1929: 18-19).

Two methodologies coexisted in Wölfflin's theory of artistic development: the first is the awareness of the inescapability of historical interpretation, its tentative nature, and the resultant circular understanding; the second is the search for regulating principles by objective, scientific means. A fundamental question of the compatibility of the two is what Wölfflin leaves us. Can a theory remain within the hermeneutic circle of understanding and the limits of its historical conditions and still articulate laws that transcend the results of interpretation and understanding?

Wölfflin leaves us with a paradox which is of great relevance today. The discipline of art history has never really understood or confronted the problem. The academy, particularly in the United States, turned its back on the theory in Wölfflin's work, and focussed on the empiricism.[1] Wölfflin has been seen as a positivist to reify the normal practice of art history. In recent years, this attitude has begun to change.

Wölfflin created an interesting synthesis out of the then current aesthetic theories, artistic practice, perceptual psychological theories, philological and philosophical theories. The unity he achieved was so convincing that until recently no one has been aware of the sources of his ideas. His legacy to art history was a method of describing art of different periods and explaining change in art. His legacy to intellectual history may be far richer, for hidden in his sources are the history of intellectual thought

---

[1] Often this empiricism was viewed as a positivist point of view (Read 1968). Christine McCorkel, "Sense and Sensibility" described the normal American interpretation of Wölfflin as a positivist and subscribed to it (McCorkel 1975: 35-50).

in the late nineteenth century when new disciplines began to achieve independence.

Bibliography

Böckh, August (1886) *Encyklopädie und Methodologie der philologischen Wissenschaften*, B. G. Teubner, second ed., Leipzig.

Couzens Hoy, David (1978) *The Critical Circle: Literature, History and Philosophical Hermeneutics*, University of California Press, Berkeley.

Dilthey,Wilhelm (1883) *Einleitung in die Geisteswissenschaften*, Duncker and Humblot, Leipzig.

Ermarth, Michael (1978) *Wilhelm Dilthey: The Critique of Historical Reason*, University of Chicago, Chicago and London.

Fawcett, Trevor and Phillpot, Clive (eds.) (1976) *The Art Press*, The Art Book, London.

Gadamer, Hans-Georg (1975) *Truth and Method*. Sheed and Ward, London.

Gantner, Joseph (ed.) (1948) *Jacob Burckhardt und Heinrich Wölfflin: Briefwechsel und andere Dokumente ihrer Begegnung, 1882-1897*, Benno Schwabe, Basel.

Gombrich, E. H. (1979) *The Sense of Order*, Cornell University Press, Ithaca, NY.

Habermas, Jürgen (1971) *Knowledge and Human Interests*. Beacon, Boston.

Habermas, Jürgen (1979) *Communication and the Evolution of Society*, Beacon, Boston.

Hart, Joan (1981) *Heinrich Wölfflin: An Intellectual Biography*, Dissertation, University of California, Berkeley, CA.

Hart, Joan (1982) "Reinterpreting Wölfflin: Neo-Kantianism and Hermeneutics," *Art Journal*, Vol. 42, No. 4, 292-300.

Hart, Joan (1984) "Some Reflections on Wölfflin and the Vienna School," in: *Wien und die Entwicklung der kunsthistorischen Methode* of *Akten des XXV. Internationalen Kongresses für Kunstgeschichte*, Band 1: Sektion 1, Hermann Böhlaus, Wien.

Kleinbauer, Eugene W. (1971) "Introduction," in: *Modern Perspectives in Western Art History*, Holt, Rinehart and Winston, Inc., NY.

Kultermann, Udo (1966) *Geschichte der Kunstgeschichte*, Wien-Düsseldorf.

McCorkel, Christine (1975) "Sense and Sensibility: An Epistemological Approach to the Philosophy of Art History," *Journal of Aesthetics and Art Criticism*, XXXIV, 33-50.

Palmer, Richard E. (1969) *Hermeneutics: Interpretation Theory in Schleiermacher, Dilthey, Heidegger and Gadamer,* Northwestern University, Evanston.

Panofsky, Erwin (1964) "Das Problem des Stils in der bildenden Kunst", reprinted in: *Aufsätze zu Grundlagen der Kunstwissenschaft* first published in 1915, Berlin.

Paret, Peter (1981) *The Berlin Secession: Modernism and Its Enemies in Imperial Germany,* Harvard University Press, Cambridge.

Read, Herbert (1968) *"Introduction"* to *Wölfflin's Classic Art,* Phaidon, London.

Ringer, Fritz K. (1969) *The Decline of the German Mandarins,* Harvard University Press, Cambridge, MA.

Ringer, Fritz K. (1977) "Higher Education in Germany in the Nineteenth Century," *Schule und Gesellschaft im 19. Jahrhundert: Sozialgeschichte der Schule im Übergang zur Industriegesellschaft,* Ulrich Hermann (ed.), Weinheim and Basel, Beltz, 332-347.

Rorty, Richard (1979) *Philosophy and the Mirror of Nature,* Princeton University Press, Princeton.

Sayce, A. H. (1874) *The Principles of Comparative Philology,* Trübner, London.

Schlosser, Julius von (1934) "Die Wiener Schule der Kunstgeschichte," *Mitteilungen des Österreichischen Instituts für Geschichtsforschung,* XIII (Erganzungen).

Willey, Thomas E. (1978) *Back to Kant: The Revival of Kantianism in German Social and Historical Thought, 1860-1914,* Wayne State University Press, Detroit.

Wölfflin, Eduard (1933) *Ausgewählte Schriften,* Gustav Meyer (ed.), Dieterich, Leipzig, 330.

Wölfflin, Heinrich (1929) *Kunstgeschichtliche Grundbegriffe,* F. Bruckmann, Munich, 7th ed.

Wölfflin, Heinrich (1946) *Kleine Schriften (1886-1933),* J. Gantner (ed.) Benno Schwabe, Basel.

What is Cultural History? Troels-Lund and the Controversies among German Historians in the 1880s.

Bjarne Stoklund

The aim of this paper is to throw some light on the process by which the field of history took shape and was delimited at the end of the 19th century. I will do so by looking at a German academic discussion of the following questions: What is proper history and what is not? What part of historical human reality should be chosen by the historian as his field of study, and what may be left out?

In the search for understanding of today's situation in the humanities one is very often taken back to the intellectual developments in Germany up to and around the turn of the century. The German universities of the 19th century were indisputably the leading ones in Europe in this field, and must therefore take a significant portion of this credit - or responsibility - for the directions the humanities took in the following century.

What was characteristic of developments in the German universities of the 1880s was on the one hand the *professionalization* of research in the humanities, and on the other the *disciplinary differentiation* that went hand in hand with it. A large number of more or less independent disciplines were distinguished, each taking its own sector of culture or society as an object of more detailed investigation. What they all had in common was a consensus that the main issues were of a *genetic-historical* nature. For 19th-century scholars, to understand a phenomenon was to lay bare its origin and development. Any complex of problems could only be dealt with in the historical context. There was an intimate connection between these main approaches to the problems and disciplinary specialization itself, which became possible precisely because the diachronic relations were felt to be more significant than the synchronic ones. So research on literature became the *history* of literature; research on art became the *history* of art; research on religion the *history* of religion, and so forth.

It was only natural that the discipline of *history* itself assumed a central position among the rich flora of historicizing specialized subjects. The discipline came well equipped for this role, first and foremost with the tools of professionalization constituted by *source criticism*, developed by Niebuhr and Ranke in the first part of the century, and which were to remain the be-all and the end-all for historians for more than a century.

Questions of the nature of historical research, its limits and its methodology were capable of arousing vehement feelings at the close of the last century. The school of historical thought that later became the dominant trend-setting one only succeeded after a number of controversies where divergent views were at loggerheads. It is worth turning back today to the polemics of these years, as the issues discussed and arguments used have remained surprisingly topical. They may even be said to have gained new interest because of the general movement in recent years in the field of historical research towards social and cultural history, under banners such as *historical anthropology, Alltagsgeschichte* and *histoire des mentalités*. [1]

The decisive battle among the German historians was the one associated with the name of *Karl Lamprecht* and concerned both the proper field of study for the historian, and his theoretical premises and methodology. Should historians content themselves with the study of *das Einmalige* - the singular historical phenomenon- within a Romantic/Idealist tradition? Or should they - as the positivistically-orientated Lamprecht would have it - move towards a collective history, a form of historical sociology partly inspired by social psychology? It was the over-ambitious Karl Lamprecht[2] who lost this *Methodenstreit*, and the result

---

[1] A good discussion of these more recent approaches seen in relation to the older tradition of cultural history in Germany can be found in Thomas Nipperdey, "Kulturgeschichte, Sozialgeschichte, Historische Anthropologie", (Nipperdey 1977).

[2] Karl Lamprecht (1856-1915) won recognition with *Deutsches Wirtschaftsleben im Mittelalter I-IV* (1885-86). His principal work, *Deutsche Geschichte*, which appeared from 1891 on in no less than 18 volumes, was a more ambitious than successful effort to write German social and cultural history. he gave an account of his methodology of cultural history in "Was ist die Kulturgeschichte? Beitrag zu einer empirischen Historik", in *Deutsche Zeitschrift für Geschichtswissenschaft*, N.F.I 1896-97. The dispute is discussed by, among others, Oestreich 1969 and Viikari 1977.

was that the historicist school became firmly ensconced both in Germany and a number of its neighbouring countries.⁻

It is not the Lamprecht battle, however, that we will concern ourselves with here, but its precursors in the 1880s. These *Vorgefechte*, as they have been called (Kuczynski 1984, cf. Oestreich 1969) were less a matter of methodology than of demarcation: should history primarily be the history of the state - political history - or should it be the history of culture in the broadest sense? This dispute involved two main combatants. One was the Prussian Professor of history *Dietrich Schäfer* (1845-1929), a specialist in Hanseatic and Scandinavian history, a supporter of Bismarck and later a passionate advocate of Germany as a great naval and colonial power. His opponent was the southern German cultural historian, later Professor of Political Economy, *Eberhard Gothein* (1853-1923).

But another major figure was also involved in the dispute as a sort of *Prügelknabe*. He was not mentioned by name, but was in fact the Danish cultural historian *Troels Troels-Lund* (1840-1921) who was engaged at the time on his great work *Dagligt Liv i Norden i det 16. Århundrede* ("Everyday Life in Scandinavia in the 16th Century"). Troels-Lund did not take any direct part in the German dispute, but later in a small work in Danish gave his answer to Dietrich Schäfer and stated his views on cultural history and its tasks.

The debate opened with a lecture by Dietrich Schäfer inaugurating his professorship at Tübingen in 1888. Schäfer had entitled his lecture "Das eigentliche Arbeitsgebiet der Geschichte" (Schäfer 1888) and began by dealing some swinging blows to the more peripheral areas of historical research:

Der Ausdruck Kulturgeschichte ist in unseren Tagen außerordentlich populär geworden. Schon die buchhändlerischen Unternehmungen, denen er als Aushängeschild dienen muß, zeigen, daß er sich großer Beliebtheit erfreut. Man glaubt vielfach, mit ihm in eine neue Ära der Geschichtsschreibung eingetreten zu sein, aus den alten Haupt- und Staatsaktionen heraus auf die Bühne des täglichen Lebens. Man meint, das Interesse von dem Hohen und Höchsten dieser Welt hinweg auf die breiten Masse des eigentlichen Volkes gelenkt zu haben und lenken zu sollen.

And what interests this type of research, says Schäfer, is not people's political or religious convictions and passions, but their daily habits and needs:

> ...nicht jene geistigen und sittlichen Regungen, die als göttliches Erbteil im Menschen leben, sondern die Formen der Befriedigung jener niedrigeren Bedürfnisse, die im animalen Teile des menschlichen Seins ihren Ursprung haben (Schäfer 1888: 265-266).

This kind of history, continues Schäfer, is particularly to be found among small peoples who in these times of great powers have no real national functions to perform - or who have voluntarily resigned from performing them. Instead, the scholars among such peoples have addressed themselves wholeheartedly to "Erforschung des Kleinlebens bis hinein in seine minutiösesten Details". And then comes the example to be shunned:

> Im skandinavischen Norden erscheint ein umfassend angelegtes Werk, das die Geschichte Dänemark-Norwegens am Schlusse des 16. Jahrhunderts behandelt. Bisher hat es sich nur mit der inneren Geschichte der beiden Länder beschäftigt, auf der später die äußere aufgebaut werden soll; aber es umfaßt doch schon eine Reihe von Bänden. Nicht weniger als ein voller Band ist allein der Geschichte des bäuerlichen und bürgerlichen Wohnhauses gewidmet, und der Autor erzählt in behaglicher Breite oder bemüht sich wenigstens zu erzählen (denn in vielen Fällen weiß man nichts davon), wie Tisch und Stuhl, Schrank und Bett, Topf und Tiegel sich gestaltet haben.

This work - of doubtful, or at least of extremely limited value - had won acclaim and praise not only at home but also in Germany, where it had appeared in an abridged form; for in Germany, too, the history of culture was rearing its head, although the situation was not as serious as in Scandinavia and the Low Countries:

> Auch bei uns sind die Kräfte einzelner und ganzer Vereine nicht selten dem beschränktesten historischen Kleingewerbe dienstbar geworden, und die Historiker der Zunft, die Vertreter des Faches an Hochschulen und Archiven, werden wohl hier und da als Männer charakterisiert, die ihre Zeit nicht verstehen. Kulturgeschichte ist auch bei uns in weiten historisch interessierten Kreisen die Losung, mit der man

neuen Erfolgen auf dem Gebiete geschichtlicher Erkenntnis und Darstellung entgegen zu gehen hofft (Schäfer 1888: 266).

I have quoted Dietrich Schäfer at such length on this point because his choice of words leaves one in no doubt that he has nothing but profound contempt for the kind of cultural history practiced by Troels-Lund and like-minded historians. Nor is he in any doubt about what is and ought to be the primary task of the historian: the study of the state, its emergence and its growth. That this was how it had been hitherto was demonstrated by a review of German historiography since the Middle Ages:

> Auch fernerhin wird es die Aufgabe des Historikers sein, den Staat zum Verständnis zu bringen, seinen Ursprung, sein Werden, die Bedingungen seines Seins, seine Aufgaben. Hier war, hier ist, hier bleibt der einigende Mittelpunkt für die unendliche Fülle der Einzelfragen, die historischer Lösung harren (Schäfer 1888: 279-280).

Such a view leaves no room for real cultural history; for the aspects of human existence it concerns itself with only assume meaning when seen in relation to the state. Recently, Schäfer continues, we have also witnessed the emergence of a number of new, specialized, historically-orientated disciplines: the history of the church; of the law; of language; of literature; and of art. Each of them sheds light on an aspect of human culture, and these are indispensable ancillary disciplines for historical research proper. Even a cultural history dealing with the phenomena of everyday life has some use, but its concerns are of a very limited kind; and it must not assume a position to which it is not entitled. This is a tendency rather to be opposed than encouraged (Schäfer 1888: 284).

With this sharp formulation of the task of historical research Schäfer had now thrown down the gauntlet to historians of culture. The only scholar whose work had been referred to specifically - although without mentioning names - was Troels-Lund; and in the first instance, as mentioned above, he remained silent. But there were others in Germany itself who were stung into responding to Schäfer. Their leading spokesman was Eberhard Gothein, who had qualified in 1878 with a thesis on "Politische und religiöse Volksbewegungen vor der Reformation", and had published a work in 1886 on "die Kulturentwicklung Süditaliens". This had won the approval of Jacob Burckhardt and given him an eminent position among cultural historians.

Gothein's answer came in 1889 in the form of a small book with the title *Die Aufgaben der Kulturgeschichte* (Gothein 1889). Here we find a very self-aware cultural history striving for a place as the natural meeting-point for the humanistic disciplines. The book has the following introductory remarks, which almost sound like a modern interdisciplinary manifesto:

> Aufstrebende Wissenschaften bedürfen und begehren keine ängstliche Abgrenzung ihres Arbeitsgebietes. Ihre Kraft beruht eben darauf, daß sie die engste Beziehung zu allen benachbarten Wissenschaften bewahren, daß sie die in jenen beschäftigten Forscher zur Mitarbeit aufrufen...ihre Aufgabe ist es, althergebrachte Abgrenzungen zu zerstören und durch bessere Kombinationen an die ursprüngliche Zusammenge-hörigkeit und Einheit aller Wissenschaften zu erinnern.

Philosophy had played such a role as centre and catalyst for the other sciences in the eighteenth century. But now philosophy itself had appointed the study of history as its successor by indicating that the development of the human spirit was one of the most important objects of research. Elsewhere in the book Gothein expresses the same idea in slightly different terms:

> Die Wissenschaft vom menschlichen Geiste ist nur *eine*. Fassen wir sie nach ihren gleichbleibenden Grundlagen auf, so nennen wir sie Philosophie, suchen wir die Wandlung und Entwicklung ihres Gegenstandes zu erkennen, so heißt sie Kulturgeschichte. *Tertium non datur* (Gothein 1889: 49).

We have seen that for Dietrich Schäfer the object of cultural history was what is left over when the humanistic disciplines have each taken their slice of the cultural pie - or in Gothein's words:

> Wenn jede von ihnen ihren Anteil in Anspruch nimmt, dann bleiben der vermeintlichen Kulturgeschichte nach dem Verluste des entlehnten Putzes nur ihre eigenen dürftigen Federn: die Betrachtung der äußeren Formen und Hilfsmittel des alltägliches Lebens (Gothein 1889: 5).

Yet it is quite another kind of cultural history Gothein has in mind: a central historical discipline to which everything else, including political history, must be subordinated. For such a cultural history to exist, it must cover a field that is at once its own and yet combines the activities of the other disciplines.

Gothein is in no doubt about the nature of this field:

> Dieser Gegenstand wird *die allgemeine Kultur einer Epoche* genannt werden. Sie ist noch etwas anderes als die Gesamtsumme aller wirtschaftlichen Leistungen, Rechtsbildungen, religiösen Meinungen, wissenschaftlichen Entdeckungen und künstlerischen Gestaltungen, sie besteht, in nicht mehr und nicht weniger als in den gemeinsamen, unter sich wieder zwiespältigen und ringenden Richtungen des Geisteslebens. *Kulturgeschichte in ihrer reinsten Form ist Ideengeschichte.* Aber nicht bloß die Idee, wie sie im schöpferischen Geiste in die Erscheinung tritt, sondern auch die Art, wie sie sich langsam vorbereitet, die Weise, wie sie sich ausbreitet, die Bedingungen, unter denen dies geschieht, die Einschränkungen und Erweiterungen, die sie erfährt, die Umgestaltung der wirklichen Welt, die sie verursacht - *das alles zu entwickeln, ist Aufgabe der Kulturgeschichte.* Darum wählt sie auch gern die *Methode der Analyse*: sie führt Ereignisse auf Kräfte, Kräfte auf Ideen zurück (Gothein 1889: 49-50, my italics).

For Gothein the great ideal for a historiography of this type was Jacob Burckhardt. In his "Die Kultur der Renaissance in Italien" (1860) Burckhardt had precisely marked out the fundamental thoughts and ideas of an epoch. This was not a matter of the history of art, literature or religion, but of the point where all these meet, and of where they come from: mankind, society. Thus develops, before our very eyes, that "modern culture" which we still call our own (Gothein 1889: 52).

If Dietrich Schäfer wanted a verbal duel between political and cultural history, he should - in Gothein's view - have chosen a worthier opponent: Jacob Burckhardt would have been an obvious choice; another possibility was Gustav Freytag, whose "Bilder aus der deutschen Vergangenheit" had some of the same qualities. Unfortunately, he had chosen the wrong opponent:

> D. Schäfer betrachtet und bekämpft als gegenwärtige Kulturgeschichte nur jene traurigen Reste, die er ihr in Zukunft übrig lassen will. Als ihre typische Gestaltung führt er ein Werk an, in dem ein ganzer Band der Geschichte des skandinavischen Hauses gewidmet ist. Er stempelt also zunächst einen zwar

nicht unnötigen aber untergeordneten Zweig der Kulturgeschichte ausschließlich mit ihrem Namen... (Gothein 1889: 5-6)

So, despite their profound differences over the "proper object" of
historical studies, the two German historians concur in their assessment of Troels-Lund. Neither of them will defend a cultural
history that only concerns itself with everyday trivia.[3]

Is the judgement passed on Troels-Lund here a just one? We
will look at this in more detail in the following; but let us first
make it clear that what is at issue in the learned German dispute
is only a small section of the work - and certainly not the most
successful one at that. In 1882 Troels-Lund had published the
second and third volumes of his work in a German adaptation
entitled "Das tägliche Leben in Skandinavien während des
sechszehnten Jahrhunderts. Eine kulturhistorische Studie über
die Entwicklung und Einrichtung der Wohnungen" (Troels-
Lund 1882). No more volumes appeared in German, and Eberhard Gothein's knowledge of the work was presumably limited
to this publication.

This was not the case with Dietrich Schäfer, who knew Danish
and could therefore follow the work volume by volume as it
appeared. But this only confirmed his negative assessment.
When Schäfer's lecture was reprinted in 1913 he added in a foot-
note: "Zum Abschluß ist es auch heute noch nicht gekommen;
1901 ersichten den 14. Band, alles allein über die tägliche
Geschichte".

When, in the fourth volume of the German work "Geschichte
von Dänemark" (Schäfer 1893) Dietrich Schäfer repeated his
derogatory views on *Dagligt Liv i Norden*, Troels-Lund was at

---

[3] Dieter Schäfer continues the discussion in a reply to Gothein two years later
(Schäfer 1891), which elaborates on his point of view, but does not contribute
new arguments. It is worth mentioning, however, that in his renewed survey of
the historical field in the broadest sense he now also refers to general
ethnology. Subjects such as *Ethnologie* and *Völkerpsychologie* are interesting
and promising, according to Schäfer; but can they be called *historical*? The
answer is no: "...solange Völker ein Herdenleben führen, sind sie selten ein
Vorwurf der Geschichte, so wertvoll die Kenntnis ihrer Zustände für den
Ethnologen und Anthropologen auch sein mag. Kulturgeschichte dieser Richtung
hat sich losgelöst von der Geschichte oder ist vielmehr, abgesehen von dem
Namensklange, nie mit ihr verbunden gewesen". It might be added that it was
from this general cultural history in Germany that Edward Tylor in his
Primitive Culture (1871) borrowed the specifically German meaning of the word
culture and changed it into an international scholarly term (cf. Kroeber &
Kluckholm 1952).

last provoked into a response - the small work "Om Kulturhistorie" ("On Cultural History") (Troels-Lund 1894). The work is a successful and vigorous contribution to the polemical genre, written with an exquisite mixture of genial charm and mordant irony.

What had provoked Troels-Lund was not so much Schäfer's belittling of his work as the fact that he saw *Dagligt Liv i Norden* as the expression of a small, weak people's national resignation. Troels-Lund, who had campaigned enthusiastically in the 1880s for a strong Danish defence, felt this to be a palpable injustice, and therefore attempted in his book to show that *Dagligt Liv* was in fact a contribution to the national revival after the military defeat in 1864. Moreover, Troels-Lund demonstrates a keen eye for the social and political roots of the historian - both as regards himself and the two German scholars, whom he places accurately as representatives of a "Greater German centralizing trend" and "South German separatist tendencies" respectively.

Nor did Troels-Lund have any illusions that the historian can content himself with describing "wie es eigentlich gewesen". Historians make a conscious selection from the many details in the sources:

> We gather these into a picture, form an ideal world where - staying as true to the facts as we can - we let the past repeat itself, but in a new way, with foreshortened lines between cause and effect, holding up what we believe to be the essential features in a light clearer than perhaps appeared to the contemporaries of the events described... But as the historian thus reproduces the past, he also paints unawares a picture of himself (Troels-Lund 1984: x).

Troels-Lund does not agree with Dietrich Schäfer in considering political history the proper field of historical studies. More important in his opinion is "the history of culture in the broadest sense - man's growing knowledge and cultivation of himself as well as of the earth on which he has been placed." Political history is indispensable, yet it is doomed to yield eventually to cultural history - or, in his own characteristically graphic words: "Sandbags full of lines of kings, acts of war and terms of peace will be emptied; and the old political history will shrink down to the dry expression 'this or that century'" (1894: xviii).

But for Troels-Lund cultural history could not be reduced - as for Gothein - to the history of ideas. For him the material changes in which his times had been so rich were a very important part of that history - or in another of his telling turns of phrase: "I concede my frailty: I would willingly trade my knowledge of several Prussian Dukes for that of the glass lamp chimney; and even forego an abundance of Electors for the use of matches" (1894: xxii).

As for the reflections that crystallized in *Dagligt Liv*, one of the things Troels-Lund has to say in his little work is this:

> The vantage point I was to choose was the sixteenth century - like our own age a dawning, a budding time of transition. Yet the view was not to be restricted to that alone: the eye would follow where the occasion beckoned, both as regards what went before and what followed after. The subject was to be the simplest, most basic conditions of life, those common to all: dwellings, food, clothing, birth, marriage, life and death - "everyday life", as I have called it... (1894: xxviii).

As for Dietrich Schäfer's distinction between the "spiritual and moral impulses that inhabit mankind like a divine heritage" and the things that "belong to the animal part of human nature", Troels-Lund counters this with a view of life as a unity:

> The existence in which we find ourselves, the life we live as human beings, is irremediably bound up with the need for sustenance, dwellings and clothing, and circumscribed by forms - birth, marriage, cohabitation and death (1894: xxxv).

Troels-Lund has little more to say in his small book on the aims of, and ideas behind, his major work on "Everyday Life in Scandinavia". To understand more, one must go to the work itself. A few words on its genesis would not be out of place. *Dagligt Liv i Norden i det 16. Århundrede* appeared in fourteen volumes between 1879 and 1901.[4] The work did not originate in a fully pre-

---

[4] T. Troels-Lund: *Danmarks og Norges Historie i Slutningen af det 16. Aarhundrede* ("The History of Denmark and Norway at the End of the 16th Century", Vols. 1-14 (Copenhagen 1879-1901)(from Vol. 2 on with the subtitle *Dagligt Liv i Norden i det 16. Aarhundrede* - "Everyday Life in Scandinavia in the 16th Century" - which became the main title from the 1903 edition on).
Vol. 1 (1879) Land og Folk / Land and People
Vol. 2 (1979) Bønder- og Købstadsboliger / Rural and Urban Dwellings
Vol. 3 (1880) Herregaarde og Slotte / Manor Houses and Castles
Vol. 4 (1882) Klædedragt / Dress
Vol. 5 (1883) Fødemidler / Food

meditated concept, but grew slowly, changing direction and being pruned down as it grew. It thus took form through a gradual process. In reality it was never quite finished: despite its huge size it is still no more than a torso.

The first volumes were meant as a general introduction to a historical work on the Dano-Norwegian state at the end of the 16th century. There was nothing novel in this plan: other historians before Troels-Lund had similarly introduced major expositions with a general description of the state of society. It was the further realization of the plans that made Troels-Lund's work into something previously unheard-of.

After describing the cultural landscape and the ethnic groups in the first volume, in the second Troels-Lund gets to grips with housing conditions in town and country; and with this volume he launches the title that was to be that of the whole work - *Dagligt Liv i Norden i det 16. Aarhundrede* (Everyday Life in Scandinavia in the Sixteenth Century). As late as the tenth volume, however, he was still - in an overview - maintaining that he would follow the original plan of also dealing with political history: the balance had simply shifted. Of the total 23 volumes planned at the time, seven were to deal with political history. Of the other sixteen, two were to be concerned with the social hierarchy and occupations. Troels-Lund has been criticized for only depicting the people "in festive dress", but the outline plan shows that he also intended to deal with people at work.

I have previously suggested that the first volume of the work, on material conditions, is the weakest as a whole. Troels-Lund had none of the first-hand knowledge, for example, that travel and fieldwork could have provided. He was happiest at home at his desk; and the first volumes are not wholly free of a certain amount of armchair theorizing. But another reason why the

---

Vol. 6 (1884) Hverdag og Fest / Weekdays and Festivals
Vol. 7 (1885) Aarlige Fester / The Yearly Cycle
Vol. 8 (1887) Fødsel og Daab / Birth and Baptism
Vol. 9 (1888) Trolovelse / Betrothal
Vol. 10 (1890) Forberedelse til Bryllup / Wedding Preparations
Vol. 11 (1891) Bryllupper / Weddings
Vol. 12 (1895) Ægteskab og Sædelighed / Marriage and Morality
Vol. 13 (1898) Livsbelysning / The Ambience of Life
Vol. 14 (1901) Livsafslutning / The End of Life
Troels-Lund and his work on cultural history have been discussed by Knud Fabricius (1921), Ellen Jørgensen (1943), Erik Kjersgaard (1968) and Bjarne Stoklund (1981).

contents of these volumes dated so quickly is that fields like housing and clothing were sub-areas in the study of cultural history that became objects of research during these very years - for example at the newly-fonded "folk museums".

With the sixth and seventh volumes the character of the work changes in several ways. The empirical basis now improves, not least because Troels-Lund could now draw on an extensive excerpt collection made by a pioneer in the study of cultural history, the librarian (later the Reverend) Nicolai Jacobsen.

At the same time people themselves come more directly under the spotlight of cultural history. From the eighth volume to the end of the work description centres on the crucial situations of life: birth and childhood, sexuality and marriage, death and burial. Here one feels that Troels-Lund is on home ground. His approach to cultural history was not - as for so many others - through things, but through people. The actual spur to his interest in the study of the everyday life and conditions of the people of the past was a collection of aristocratic letters from the 16th century that he had chanced upon during his work as a young graduate in the State Archives; these opened his eyes to the opportunities offered by personal documents like letters and diaries for in-depth study of the lives of people in the past.

The life cycle is concluded with Volume 14, which deals with death and burial. But before this he had inserted a thirteenth volume on "The Ambience of Life" - not included in the plan, and interrupting the treatment of the life cycle. Contemporary reviewers had often criticized Troels-Lund by saying that "Everyday Life", with its huge scale and many details, tended to fall apart at the seams and was having trouble arriving at a synthesis. He himself was undoubtedly aware that he was having difficulty fulfilling the final goal of the plan: "to see clearly how the great enigma we call life appeared to the people of those times."

The volume on "The Ambience of Life" was meant to compensate for this. In the introduction he says:

> We have seen how the generations of men who lived in the north in the 16th century came into the world, were housed, were clothed, ate and drank, married and lived as man and wife. Before we commit them to earth and thus conclude our consideration of the "everyday life" of those times, there remains one great question to be investigated, the most important and difficult of them all. "Everyday" life has been appre-

hended merely in its externals, all the while we have gained no impression of the light in which life appeared to those generations - the shades that coloured all things, that coloured the act of living itself...

It is a matter of capturing "this most elusive quintessence of history", the "interplay between each passing generation and its external conditions; the fragrance and colouring that emanated from these for the age itself while it yet lived, before it was dried out and pressed between the pages of history." The volume would attempt to show "how that age saw, not how we see" (*Dagligt Liv*, 6th ed., Vol. 7: 9-10).

If, after such a preamble, the modern reader expects a synchronic cultural analysis or an experiment in the history of mentalities, he will be disappointed. What Troels-Lund does is something quite different:

Like any growing thing, the conception of life of those times can only be grasped from below upwards; by an investigation of why and how it came to be so. Like plants, these feelings and frames of mind have thrust their way up from the soil in the course of millenia, borne up by the thinking of many generations; they have assumed form and drawn sustenance from the impressions not only of their own, but of previous ages. Like plants, they can be understood only in their entirety, not from their mere severed tops (6th ed., Vol. 7: 10).

When Troels-Lund here - and elsewhere - speaks of wholeness, he means it in the diachronic sense - a process of development whose metaphorical expression incidentally reveals its affinity with the organicist thinking of Romanticism. The idea that a phenomenon can only be explained by being placed in a historical context is part, as mentioned before, of the conceptual framework that bound scholars of many different schools together in the 19th century. In accordance with his outlined programme, Troels-Lund attempts in the volume on "The Ambience of Life" to trace out some characteristic features of the 16th century - emancipation from the power of the church, love of nature and fascination with demonology - in a long evolutionistic view from the earliest cultures on. At the same time the book became a kind of personal commitment to the joyous, optimistic belief in development that could still find a sympathetic response in the last decades before the Great War. The book sold better than any of his others; but by scholarly criteria this volume is one of the weakest in the whole work.

In general, the idea of evolution plays a major role in "Everyday Life in Scandinavia". It emerges at many points in the work and makes its presence felt even in the treatment of the detailed elements of material culture.

The idea of evolution is one of the two theoretical premises in Troels-Lund's work; the other - only implicitly present - is a concept of culture which sees it as the sum of the elements it contains. Thus, understanding culture becomes a matter of investigating its particularities and then placing them in a diachronic context.

In these theoretical presuppositions there was little help to be had for the cultural historian who wished to grasp the cultural character of a specific period. Faced with this task, Troels-Lund, too, was obliged to trust his hermeneutic intuition, and this is also what he does at several points in the work. One is also aware at such points of his conscious striving to break free of his own values and norms and those of his time and understand Renaissance man on his own terms.

As an example let us take his treatment of drunkenness in the volume on food and drink. Here Troels-Lund arrives hesitantly at the conclusion that sixteenth-century man has a completely different view of this phenomenon from us. And to this realization he adds the following lines, which reveal a significant side of his view of cultural history:

> We stand here at one of the points where spirit and matter meet, where the connection between the foodstuffs and purely mental manifestations of life emerges. To term the one a mere cause, the other a mere effect, will not do. Both are aspects of one and the same thing, elements in the common cycle of life. To that mode of life there corresponded, as condition and result at once, conceptions of the domain of morality and hygiene far different from our own.

"If we press our investigations further," continues Troels-Lund,

> It emerges that beneath a light veneer of obligatory disapproval of drunkenness lay quite another view... The true, honest feeling was one of anything but regret. In this view *to be in one's cups was an act of merit*; a sign that the person in question was a fine, amiable, upstanding individual. To aban-

don oneself so was the expression of a good nature; to be drunk often the touchstone of a man of worth, a man made of the right stuff, a man without duplicity (6th ed., Vol. 3: 237-239).

Yet Troels-Lund goes a step further, and attempts to see drunkenness and its accompanying behaviour as an element in a total pattern of life typical of the time:

We have considered the concepts in the areas of morality and public health that corresponded to the mode of life of the sixteenth century. We may venture a step further and strive to clarify the *frame of mind and the whole view of life* that must have been natural in circumstances such as these.

The features that typify the age, and that must be seen in relation to drunkenness, are according to Troels-Lund its *violence*, with its many excessive killings; its *fertility* and *immorality*, and that *excessive indulgence in pleasure* "which cannot but astonish a more temperate age". "As a unifying feature for this violence, anger and excess, there was finally, as a common characteristic of the age, a unique *infatuation with reality*, an almost headlong *total involvement* in everything one did (6th ed., Vol. 3: 246-254).

Passages like the one quoted here grant us, much more so than Troels-Lund's own deliberations, insight into what he wanted to achieve with his work, and why it has remained relevant.

*   *   *

Let us make a halt here to summarize. In the last decades of the nineteenth century history and the other humanist disciplines took form in the course of a number of exchanges of opinions, particularly in Germany. We have attempted to shed some light on this by discussing one of these scholarly feuds, where the main actors represent three different concepts of the domain of history, and in particular of the role of cultural history.

For Dietrich Schäfer true history was political and its object was the growth of the power of the state. A number of specialized disciplines each had the task of dealing with its own sector of the history of culture, and in his view all that was left over for a specific cultural history were certain inferior aspects of the human condition. One could say with some justification that Schäfer represents the school that prevailed. In Germany and many of its neighbouring countries university research was for several generations predominantly political, while cultural history was banished to museums and other institutions outside the university. Here an often very specialized type of cultural history was studied, while the great syntheses were left to loners whose research was regarded with some scepticism. In Denmark cultural history was not admitted to the University of Copenhagen until about 1960, and even then with a very narrowly defined working area - popular material culture. It was not until 1971 that a more comprehensive discipline termed *ethnology* was established.

The two other scholars involved in the dispute saw cultural history as the most important side of historical research, but conceived of it in very different ways. Eberhard Gothein represented a purely idealistic view which saw ideas as the motive forces in history and identified cultural history with the history if ideas. This was the school of cultural history founded by Jacob Burckhardt, whose foremost representative was Johan Huizinga.

Against this stood Troels-Lund's view of cultural history as the history of everyday life of human beings. He was equally interested in the material conditions of life and people's own views and interpretations of their daily lives. The two things could not be separated, but were inextricably bound up with one another. For Troels-Lund human life was a whole.

In today's growing field of cultural history one may find representatives of a *history of mentalities* who are often not very far from the idealistic views of Gothein. But we also find a *cultural history of everyday life* which may be seen as continuing the tradition from Troels-Lund. Like their great predecessor, cultural historians of this kind often run the risk of being overwhelmed by the mass of the details of daily life. But the positive lesson that modern cultural history can learn from Troels-Lund is that everyday human life must be regarded as a whole; and that material forms, and cultural norms and interpretations, are two sides of the same thing that cannot be separated and studied independently of each other.

Bibliography

Fabricius, Knud (1921) *Troels-Lund*. København

Gothein, Eberhard (1889) *Die Aufgaben der Kulturgeschichte*. Leipzig

Jørgensen, Ellen (1943) *Historiens Studium i Danmark i det 19. Aarhundrede*. København

Kjersgaard, Erik (1968) Manden og Værket, introduction to *Dagligt Liv i Norden*. 6.ed.. København

Kroeber, A.L. & Clyde Kluckhohn (1952) *Culture, A Critical Review of Concepts and Definitions*. New York

Kuczynski, Jürgen (1984) Vorgefechte im Meinungsstreit, in: *Zeitschrift für Geschichtswissenschaft* 32. Jahrg., p. 429-439.

Lamprecht, Karl (1896/97) Was ist Kulturgeschichte? Beitrag zu einer empirischen Historik, in: *Deutsche Zeitschrift für Geschichtswissenschaft* N.F.I.

Nipperdey, Thomas (1977) Kulturgeschichte, Sozialgeschichte, Historische Anthropologie, in: Schieder & Gräubig (eds.) *Theorieprobleme der Geschichtswissenschaft*. Darmstadt, p. 286-310

Oestreich, Gerhard (1969) Die Fachhistorie und die Anfänge der sozial-geschichtlichen Forschung in Deutschland, in: *Historische Zeitschrift* 208, p. 320-363

Schäfer, Dietrich (1888) *Das eigentliche Arbeitsgebiet der Geschichte*, Jena. Reprinted in *Aufsätze, Vorträge und Reden* 1, Jena 1913, p. 264f. References in text are referring to the 1913-edition.

Schäfer, Dietrich (1891) *Geschichte und Kulturgeschichte. Eine Erwiderung*. Jena. Reprinted in *Aufsätze, Vorträge und Reden* 1, Jena 1913, p. 291f.

Schäfer, Dietrich (1893) *Geschichte von Dänemark* IV. Gotha

Stoklund, Bjarne (1981) To danske kulturforskere i grænselandet mellem historie og antropologi, in: *Folk og Kultur*, p. 55-72

Troels-Lund, T. (1879-1901) *Dagligt Liv i Norden i det sekstende Aarhundrede* Vol. I-XIV, København (cf. note 4)

Troels-Lund, T. (1882) *Das tägliche Leben in Skandinavien während des sechszehnten Jahrhundert. Eine kulturhistorische Studie über die Entwicklung und Einrichtung der Wohnungen*. - Deutsche vom Verfasser bersorgte Ausgabe. Kopenhagen

Bjarne Stoklund

Troels-Lund, T. (1894) *Om Kulturhistorie*. København. Reprinted as introduction to 3. and 4. ed. of *Dagligt Liv i Norden* 1908 and 1914, (references in text to the 4. ed.)

Viikari, M. (1977) *Die Krise der "historischen" Geschichtsschreibung und die Geschichtsmethodologie Karl Lamprechts*. Helsinki.

Saussure, Structuralism and Laudan's Problem

Frans Gregersen

Once upon a time - in the beautiful Geneva by the enchanted lake - a child was born. He was named Ferdinand and inherited the noble name of de Saussure.

This fairy tale introduction is legitimated only by the fact that we cannot pretend to understand Ferdinand de Saussure if we do not perceive the magic that surrounded his fate from the very beginning. In the rest of what I shall be saying, I shall skip this altogether but I thought the chord should be struck now, in order that it might still vibrate and thus accompany the more detached views presented in what follows.

Saussure was born on the 26th of November 1857. The family was an ancient Geneva family, counting among its members numerous famous natural scientists. In the *Souvenirs* of his childhood Saussure mentions his earliest work as a childish attempt within the science of language. Published in 1978 this is the *Essai pour réduire les mots du grec, du latin et de l'allemand à un petit nombre des racines*. It was written in the summer of 1872 when Ferdinand de Saussure was not yet 15 and its sole interest for us, lies in that it shows the literally life long interest that Saussure was to take in *Reduction*. (Essai pour *réduire* ...) and in *Roots*. The one, of course, follows from the other: If you want to be able to reduce something to its essentials, this presupposes a thorough knowledge of the primitives that are the end result of the reduction process: Ergo: *racines*.

By the 14th of May 1876 Saussure had been allowed to enter the *Societé Linguistique de Paris* and shortly afterwards the young man moved to Leipzig to study. Saussure's parents thought that it would be best for him to be with his Geneva friends and as they were to go to Leipzig he could just as well go there, too. This is the version of the main protagonist himself in his *Souvenirs*, but later observers have rather pointed to the unique character of Leipzig as the headquarters of that army of young linguists, soon to be known as the Neogrammarians.

Saussure studied with the teachers of the Neogrammarians and in 1878 he published his *Mémoire sur le système primitif des voyelles dans les langues indo-européennes.*

The *Mémoire* is so famous in itself, that had Saussure not published anything else, he would still occupy a prominent place in the history of comparative linguistics. I shall not, however, focus on the contents of the book, I just want to direct your attention to the fact that it appeared in 1878 simultaneously with the preface to Osthoff and Brugmanns *Morphologische Untersuchungen.* This preface is normally taken to present the Neogrammarians' views to the public for the first time.

The Neogrammarians - I shall not say very much about them either - represent a naturalist trend within comparative Indo-European research, virtually dominant within linguistics for the next forty to fifty years. Their period marks an interest in the physiological phonetic underpinning of the so-called exceptionless sound laws that were supposed to be the net result of comparative research *auf dem Gebiete der indo-germanischen Sprachen.* Note the substitution of *germanisch* for *europäisch, Indo-germanische Forschungen* were a German specialty.

Two years later - in 1880 - Saussure got his degree from Leipzig University. By writing his dissertation to the Masters of Leipzig, Saussure at once signalled his allegiance to the comparative method and got a certificate as a trained researcher - a comparative Indo-European linguist - made in Germany.

In the summer of 1880, immediately after completing his dissertation, Saussure was off to Lithuania. Lithuanian is a very archaic language with almost no old literature, and this of course made it very popular with the Neogrammarians. Saussure does not, however, seem to have stayed there for long, the young doctor headed straight for Paris. In 1881 he was already appointed *maître de conférence de gothique et du vieux allemand* at the École des Hautes Études and this seems to have been the first time that *Indo-germanistik* was taught in France. Saussure, as you might have guessed, was an unusually talented - and an unusually successful - teacher. During the decade from 1881 to 1891, when Saussure moved to Geneva, he taught a wide range of courses to a substantial number of students. All of these courses were firmly entrenched in the comparatist tradition but we should still single out the course of 1885-86, since by Saussure's own admission the students attending that particular course were so experienced that he could touch on matters of general linguistics.

Saussure could presumably have stayed in Paris to be appointed successor to the only relevant professorship, that of Michel Bréal, but this would have meant his exchanging a Swiss citizenship for a French one. And this apparently was not what he wished. Be that as it may, Saussure moved to Geneva in 1891.

In the beautiful Geneva they had created an extraordinary professorship for him in the history and comparison of Indo-European languages. One of the more conspicuous changes must have been the number of students. In Paris Saussure had had many-and talented ones at that, in Geneva he had few but some of them were to play important roles in the history of linguistics. Two of them, Charles Bally and Albert Sechehaye, we shall return to in a little while. The number of students actually was only symbolic of the difference between the metropolis and the provincial town. From now on Saussure was to remain outside of both the German and the French mainstream of linguistics.

In 1896 Saussure was made an ordinary professor of Sanscrit and Indo-European languages but it was not until 1906 that he was as well chosen to fill the vacant chair as professor of general linguistics. It is a matter of some interest that the University of Geneva in its statutes of 1872 had a chair of general linguistics and philology. In 1873 the rabbi of Geneva, Joseph Wertheimer, was appointed professor and he stayed in office until 1906 when Saussure took over. The title professor of general linguistics and philology must have meant something else in 1872 than Saussure made out of it 35 years later.

Saussure taught a total of three courses. The first course lasted from January till July 1907, the second one lasted from November 1908 to June 1909 while the third and last one started in late October 1910 and ended in the beginning of July 1911. The first course was attended by 5 or 6 students, at the second one the number had risen to 11 and the third one was attended by 12. Although notes from only between four and six students have been preserved, we may presume that all of them took notes. This was what one did then, the tradition of the high Middle Ages of a teacher lecturing in front of a busily writing audience had not yet died.

The last year of his life, Saussure was so ill he could not teach. He died in February 1913.

The fame and influence of Ferdinand de Saussure rest firmly on his book from 1916 *Cours de Linguistique générale*. Yet Saussure never published any book, let alone prepare a manuscript for one. The book, in fact, is a tightly edited version of some scattered notes of Saussure's and the students' notes from the three courses, mainly the first and the last. The editors were Charles Bally and Albert Sechehaye, devoted pupils, but - it is to be noted - not among the students attending the courses. The *Cours* is thus an edition of the second hand student interpretation of an oral presentation and only a thorough examination and comparison of the preserved notes will lead us to what Saussure might in fact have said. It is important to understand that the bible of modern linguistics, the *Cours*, is like the New Testament in that God's words do not sound directly to us.

This complicated situation may give rise to many misunderstandings if one does not sharply distinguish between *on the one hand* the attempt at reconstructing the original Saussurean thinking. This branch of the history of linguistics is Saussurean philology, founded in 1957 by Robert Godel. In 1968 Rudolf Engler published a critical edition of *Cours* presenting more than 3000 bits and pieces of students' notes and the original text of the 1916 edition synoptically reproduced so that everyone may judge the editorial work of Bally and Sechehaye. *On the other hand*, it is of course just as legitimate to study the influence that Saussure has had via the edition of his thoughts in *Cours*. Certainly neither Bloomfield nor Hjelmslev, to name only two of the most obviously influenced persons, could have suspected that some of the thoughts in *Cours* were more Bally and Sechehaye than the master himself. They read the *Cours* as it is and so we have to do the same.

This paper is not an essay on Saussurean philology. And yet the *Cours* simply cannot be read as if time had stood still. So, although I shall concentrate on the 1916 edition of *Cours*, what I have to say will be based as well on other, newer, sources, notably the autograph manuscript on morphology edited by Robert Godel in 1969.

First of all, I want you to realize what genre this book belongs to. The title is in fact very revealing, this is a *Course in general Linguistics*. A course with a very visible and audible teacher who does everything pedagogical to explain the difficult problems for his students. This more or less explains the *style* that Saussure uses throughout. All expressions are pointed to the extent that we feel confronted with paradoxes. Or proverbs, rules, by which to live as a linguist ever after. *'Il faut'* is the usual expression. Hugo Schuchardt in his perceptive review of *Cours* drew the readership's attention to another facet of the genre:

Dazu die Form der Darstellung: der eindringliche, gebiete-
rische Vortrag des Lehrers, der jeden Widerspruch übertönen
will, auch den eigenen. Die übergrosse Sicherheit des Aus-
drucks verrät eine halbbewusste Unsicherheit in der Sache.
Saussure übersieht nichts was man einwenden könnte; er hebt
das Schwierige, Auffällige, Paradoxe hervor; schiebt aber dann
mit einer starken Handbewegung die Hemmnisse beiseite. Die
in zahlreichen Varianten auftretende Grundformel lautet: die
beiden Dinge sind schwer voneinander zu trennen, aber es ist
unmöglich sie nicht voneinander zu trennen.

Since Saussure did precisely not write down the whole of his lectures we may suppose that what has actually happened is that
while discussing with himself in the way Schuchardt has just
described, Saussure has found new solutions to old problems. But
A Course in general Linguistics is not efficient as a paradigm if it
contains contradictions. This was apparently keenly felt by the
editors and accordingly they arranged the material so that it
seems just like a deductive system, consistent and cohesive
throughout. Not only did this necessitate rearranging and editing
- sometimes they had to clarify and supplement with sentences of
their own, some of which became very famous.

Saussure's intention is to make linguistics a science. What
bothers him is that everyone else seems to take for granted that it
already is one. In a letter to Meillet as early as 1894, Saussure
expresses in the strongest possible words his disgust with the
logical shortcomings of the linguistics of the day. It is impossible
to write ten lines without using terms that have no legitimate
definition, he says. Preoccupied with the logical basis of a linguistic theory, he formulated his aim as to show the linguist *what he
actually does*. It seems then, that Saussure was not keen on establishing another linguistics but worried about the fact that the
linguist worked as if everybody knew why he was doing what he
was doing, thus committing in fact all sorts of theoretical *faux pas*
- and all the time nobody noticed.

It is normal to consider this from another angle: Instead of the
interpretation I have just given, Saussure is seen as advocating
systemic or, as it was later called, synchronic structuralist analysis,
instead of comparative historical ones, but this would be to un-
derrate the radical point of departure. In fact, what Saussure asks
is the fundamental question: What is a linguistic fact? Nothing is
easier than to proceed with business as usual: Grasp a text and
start analyzing the words in it into morphemes and roots, estab-

lish correspondences and study the words in the etymological dictionaries. Go on just do it! But is this really language we are studying, how can we be sure? If we do as the Neogrammarians would have us do and as Jost Winteler *and Saussure himself* had done, that is: 'go listen to the people talk', it soon becomes a more vexing question to find the facts.

It seems that not very many people even among educated academics know this and so it may come as a surprise to you as well: People simply do not speak sentences and they do not normally isolate single words and make pauses to mark the spaces between them, so how do we isolate the linguistic fact? The answer is: we compare signs. What Saussure did was in practice to develop the methods used by the Neogrammarians for comparison of attested languages into a comparison of forms in spoken contemporary language as well. In the comparison of forms lies the answer to our problem: if we compare the singular *fot* with the plural *foti* we conclude that the root is *fot* and the sign *zero* marks the singular, the *i* the plural. If in texts representing the speech of later ages we find *fot* contrasted with *føt* we conclude that this system of number is of a completely different kind. Here the contrast in the vowels co-occurs with the contrast in number. For brevity's sake we express this as: The language forms its plurals by internal vowel change of the word instead of suffixing *i*'s.

Let us finally take a look at a system where we have *fot* contrasting *føti*. Here the plural is signalled by an appended *i and* the change in the vowel; what we see is a process of the type that in the Nordic languages has been called *umlaut*. This last system gives us the clue to the development of the form *føt*. We could say that the *i*-ending has brought about a change in the vowel and then the *i* has been deleted. Probably this *is* what happened in some cases, but what Saussure wanted to teach us, is that at any given time the elements in these three systems were quite distinct, in short they precisely enter into separate systems. In the first system the *i* contrasts with the *zero,* in the second the *i-zero* and the *o-ø* contrast signal plural and in the last system the *ø* alone enters into contrast with the *o*. Saussure coined the term *synchrony* for this type of analysis of systems at an abstracted point in time. The contrasting analysis of successive stages he called *diachrony*.

Let us pause briefly here. We do not easily understand what Saussure did with this provocative distinction. We *do* not - *cannot* - understand this provocation because we ourselves have learned the lesson. No one seriously questions the difference between say history and sociology, but remember: times have changed and Saussure lived in a historicist society. The academics of the *fin de siècle* were no less than their fathers steeped in the classical tradition. They knew the history of their nation and the history of their language. In fact both German and Roman philology had the status of *explanatory* disciplines *vis à vis* modern usage. We say *feet because* the Anglo-Saxons said *footi* (if they did, which I do not know, but you get the idea I am sure). Saussure denied that there was anything to explain. Anglo-Saxon had one system, modern English another.

Saussure realized of course that time is not a discontinuous element so that it can be chopped up in pieces where its winged chariot has been brought to a stand still:

> En pratique, un état de langue n'est pas *un point* mais *u n espace de temps* plus ou moins long, pendant lequel la somme des modifications survenues est minime (p.142).

The point may be ten years, a generation, a century perhaps. But this definition will not do. For how do we know whether the sum total of modifications do not add up to a new system. We delimit the period by investigating when the system has changed and we describe the system by arbitrarily fixing a period. Either this is frightfully dialectic or just a good old vicious circle.

Making linguistics a science involved more than separating Indo-European comparative linguistics from synchronic analysis of living languages. First of all, linguistics had to be delimited from the neighbouring sciences. But as such, language is itself mixed up with all human activity, it is part of aesthetics, of history, philology and psychology, so what is so special about linguistics? Saussure's answer was again a methodological one: He distinguished between *langage* or *faculté de langage, langue* and *parole*.

By *parole* he means to delimit at once the individual from the social and the instantiated use of the language system from the system as such. And by *langue* Saussure correspondingly understands the system of language that lies behind the actual use, and this is social. The *faculté de langage* is the term Saussure invents to cover the ability to speak as a specific characteristic of man. Since the system does not reveal itself except through the individual *parole* we are forced to conclude that the preferred method of analysis, has to be abstraction *from* the *parole to* the *langue*. Saussure of course knew, that there could be individuals who, for some reason or other, spoke differently than the rest of the speech community, but he probably believed that the abstraction process could be corrected by studying more than one individual. To my knowledge, Saussure does not anywhere in the *Cours* speak of field methods. Perhaps he took it for granted, that we know what we want to know, about the language we want to study. In fact, he had not only worked himself in Lithuania, he had also done research on the dialect of Fribourg and owned Jost Winteler's pathbreaking study of the Kerenzer Mundart. This study is normally described as being socio-linguistics *avant la lettre*.

The practice of linguists has varied very much in this respect but to my mind it is true to say that structural linguistics suffers from what William Labov has termed the Saussurean paradox: Combining the accessible individual use of language with the not directly accessible abstract social system yields the following paradox: To study the social systems we only need an individual; to study the individual use we need a speech community.

To outsiders this seems perhaps sheer nonsense. But a crude example will show the reality of the paradox: Suppose that we want to find out how the number system of the dialect that we are studying is constituted, we cannot simply go ask somebody. The spoken language of any centralized community is heavily infected by all kinds of prejudices and norms concerning what the correct form of a word is. If for instance I ask my students to pronounce seventeen words of the type *myre*, meaning ant, vs. *myrer* ants, most of them will feel obliged to produce hypercorrect forms with very distinct syllables *my:-rø, my:-rær* but this is certainly not what they say, when I am not around. But if I, on the other hand, patiently wait for my informants to utter singular and plural forms, I shall have to wait for a long time before they produce the data that I can produce myself within an hour simply by virtue of my speaking the language. This is precisely what Saussure aptly captures with his distinction between *parole* and *langue. Parole* is organized as a chain of speech - and I might add - directed towards the end of expressing some specific content, while *langue* on the other hand is a system that does not participate in the flow of time and meaning in that way, but rather is a construction of simultaneously present correlations. Presumably it was not Saussure's intention when pointing to *langue* as the unique object of linguistics to stress the autonomy of linguistics but it most certainly was that of the editors Bally and Sechehaye.

The most often quoted sentence from the *Cours*, apart from the one on semiology is, of course, the last sentence. It reads:

> *La linguistique a pour unique et véritable objet la langue envisagée en elle-même et pour elle-même.* (Linguistics has as its sole and true object *la langue* conceived in itself and for itself).

This has, I am happy to say - no basis whatsoever in the material left from the courses. Of course the first part of the sentence is only what Saussure had stressed in the introduction, but the last part stresses what has since then been called the autonomy of linguistics as a science. The Neogrammarians had fought their war of independence against the philologists, now the linguists declared their independence in the face of all the other sciences of the humanities. This happened while the real war, the First World War, destroyed the homeland of those Indo-European comparative linguists who collaborated with students of archaeology, history and religion. The torch of Prometheus passed to France and Geneva.

In 1928, then, they all assembled for the first time, for the first time asserting in practice the autonomy of linguistics as the study of structures, not of specific languages. The Hague was the meeting place and the 1st International Congress of Linguists the occasion. It must have been a curious event, I would not mind to have been present. In The Hague, the past and the present met with each other and shook hands with the future. Baudouin de Courtenay, the Russian linguist who had been a friend and independent co-worker of Saussure's, was there, although he does not seem to have said anything, Meillet, Saussure's favourite pupil and the only one he wrote letters to consistently all during his lifetime, and Bally and Sechehaye, Saussure's self-appointed editors and heirs were there. And Hjelmslev, who was to carry through to its logical conclusion many of the tenets of the *Cours*, sat by and listened to all the speeches.

The change had come. The cultural hegemony had passed from Germany to the French speaking world and inseparably tied to this development, Saussure's ideas had become the natural basis for any worthwhile discussion of general linguistics.

The 2nd International Congress met in Geneva by the beautiful lake.

Comparative Indo-European linguistics had from its very beginnings been useful to the nation builders of Central Europe. It
created - the otherwise missing - link between, on the one hand,
the culture of classicism as this was handed down to succeeding
generations by that eternal school teacher the *magister*, and on
the other hand the new professional culture based upon the
teaching of the mother tongue. The great philologies served to
legitimate the literature of the modern languages as worthy of
replacing the all European classical tradition. And the modern
philologies were unthinkable without Indo-European foundations. Growing into a full fledged professional science during the
first three quarters of the 19th century with university chairs,
periodicals and congresses, comparative linguistics around the
late 1870s, however, broke away and tied itself to the naturalist
movement. It was thereby infused by a new energy and vitality,
but it lost the evident legitimation it had had as part and parcel of
the philologies. It is no wonder that Saussure sensed the problems with positivist naturalism in the '90s; nearly everybody as
perceptive as that did, witness Stuart Hughes' book on this
period. The activists had everywhere run into trouble, either
politically or scientifically, and a new sense of being ruled by laws
that the intelligentsia was unable to control, rose among them.
Saussure was probably the only linguist who tried to solve the
problems and it is interesting to note here that in fact *he gave it
up*. It was not until he was forced to, that he formulated his
views on the crisis of linguistics and still less had he thought of
their being published. As it happened, his ideas were in fact published and immediately had an impact. A new role for the intelligentsia emerged, the distant analyst who professionally is called
upon to unravel the hidden structure behind the accidental and
floating *parole*, read everyday reality. This meant the gradual but
total separation of linguistics from applied linguistics, a split that
is nowhere more visible than in the almost total absence of serious linguistic work on reading and writing until the late 60ies of
this century.

I now pass on to the last part of this paper, the one on Laudan's
problem. Since I cannot presuppose that everyone here is familiar with Larry Laudan's book *Progress and its Problems*, a brief
sketch of the structure of the argument is appropriate.

I hope everybody here would agree with me, when I state that
a sound theory of science at the very least should take into consideration the results of the *history* of science. In order to postu-
late criteria for choosing between theories we have to know how

the choice has actually been made in the past. Whether we would want to judge the choice an unhappy or a happy one for the future fate of the discipline, we should know about it. This presupposes that we can identify the elements of the history of science. 'Easy', they say, 'the elements are theories.' Now Laudan, following Lakatos, distinguishes theories as more local and specific statements from the more global 'research traditions'. Imre Lakatos claimed that theories follow from the axiomatic hard core of the research programme, that is a theory is seen as part of the research programme if and only if it accepts the fundamental tenets of the programme. This in fact does not allow for internal strife within a research programme or rather it limits internal disagreement to minor matters.

This point is very crucial and Laudan correctly refutes it, both theoretically and factually. It is most certainly the case, that two structuralist theories could disagree on very important points and still we would like to be able to call them both structuralist, as indeed we do. But what happens if we give up the criterion that theories must share *all* the basic assumptions, obviously they must share *some* in order for us to classify them as belonging to the same research tradition. Among the structuralists Hjelmslev accepted the dichotomy synchrony-diachrony, but the Prague School, notably Roman Jakobson, did not. Conclusion: We cannot make this distinction the cornerstone of a definition of the structuralist research tradition. We could of course start defining several structuralist traditions, but we would soon find that the difference between theory and tradition had gone. Laudan, however, chooses to say that research traditions are historical phenomena, they evolve, and consequently what is taken as the hard core of the research tradition is liable to change:

What constituted the essence of the Marxist research tradition in the late nineteenth century is substantially different from the 'essence' of Marxism a half century later. Lakatos and Kuhn were right in thinking that a research programme or paradigm always has certain non-rejectable elements associated with it; but they were mistaken in failing to see that the elements constituting this class can shift through time. By relativizing the 'essence' of a research tradition with respect to time, we can, I believe, come much closer to capturing the way in which scientists and historians of science actually utilize the concept of a tradition (p. 99f).

Saussure, then, was first the real Saussure whom nobody will ever know, and then became the Saussure of *Cours*, thereby inadvertently fathering modern descriptive linguistics only to be reinterpreted during the last two decades in the search for the authentic thinker Ferdinand de Saussure. Saussure is like a palimpsest. A new layer is added every time a generation needs a godfather.

The question is now, are all these Saussure's, and the many different models have not been described in detail for obvious reasons, are they really all of them structuralists. This is Laudan's problem: If we define research traditions broadly enough we come close to the definition that the linguists use themselves but we lose precision. On the other hand, if we define it too narrowly we lose the generalizations implicit in the term itself.

This certainly is a problem for the historiography of linguistics: If we define structuralism as a predominantly synchronical descriptive trend, preoccupied with methodological problems, it is true that the characteristic features of the 20th century as opposed to the previous century are broadly structuralist. But is that saying anything interesting about linguistics?

There is another way out: Instead of only broadening the definition of the research tradition, seen as a research tradition of some specific object, we might go even more structuralist and define the tradition as a unique way of thinking research in the humanities. This would lead us to ask questions like: What is the similarity between the way Jean Piaget thinks psychology, the way Cézanne and the cubists painted, the way Schönberg composed, the way Althusser philosophizes or the various constructivist tendencies within the arts - and the way the structuralists think language.

Another way out is to class the different structuralisms with other theoretical stances in relation to eternal epistemological dichotomies like materialism versus idealism, realism versus conventionalism or the different conceptions of the subject. This in turn might lead to defining the dynamic materialist structuralism of Roman Jakobson as belonging to another research tradition than say idealist Hjelmslevism. Whether one takes one or the other way out, the road leads straight to cross-disciplinary conferences such as this one.

Bibliography

Actes du premier Congrès international de linguistes. À La Haye (1928) du 10-15 avril 1928. Sijthoff's Leiden.

Bloomfield, Leonard (1924) Review of Cours de Linguistique générale. *Modern Language Journal* 8, p. 317-19. Repr.: *Cahiers Ferdinand de Sassure*, 21, p. 133-35, 1964.

Engler, Rudolf (1967ff) *F. de Sassure: Cours de Linguistique générale, édition critique* par Rudolf Engler, Fasc. I-IV, Harassowitz, Wiesbaden.

Gambarara, Daniele (1972) La bibliothèque de Ferdinand de Saussure, *Genava*, Nouvelle Serie 19-20, p. 319-367.

Godel, Robert (1957) *Les Sources Manuscrites du Cours de Linguistique Générale de F. de Sassure*, Droz et Minard, Genève et Paris.

Godel, Robert (1969) (ed.) *A Geneva School Reader in Linguistics.* Indiana University Press, Bloomington and London.

Godel, Robert (1973) À propos du voyage de F. de Saussure en Lituanie, *Cahiers Ferdinand de Sassure*, 28, p. 7-22.

Hughes, Stuart (1977) *Consciousness and Society*, 2nd Edition, Open University.

Jakobson, Roman (1984) La théorie saussurienne en rétrospection, *Linguistics*, 22, 2, p. 161-196, Mouton Amsterdam.

Labov, William (1975) Empirical Foundations of linguistic Theory, in: Austerlitz (ed.) *The Scope of American Linguistics*, p. 77-133, Peter De Ridder Press, Lisse.

Lakatos, Imre (1970) Falsification and the Methodology of Scientific Research Programmes, in: Lakatos and Musgrave (eds.) *Criticism and the Growth of Knowledge*, p. 91-196, Cambridge University Press.

Laudan, Larry (1977) *Progress and its Problems*, Routledge and Kegan Paul, London.

Redard, Georges (1976) Le Voyage de F. de Saussure en Lituanie: Suite et fin? *Cahiers Ferdinand de Saussure*, 30, p. 141-150.

Saussure, Ferdinand de (1872) Essai pour réduire les mots du grec, du latin & de l'allemand à un petit nombre de racines, *Cahiers Ferdinand de Saussure*, 32, p. 73-101, 1978.

Saussure, Ferdinand de (1878) Mémoire sur le Système primitif des voyelles dans les langues indo-européennes, in: *Recueil des publications scientifiques de Ferdinand de Saussure*, p. 1-268, Payot, Geneve, 1921.

Saussure, Ferdinand de (1894) Lettre à Antoine Meillet (4 Janvier 1894) *Cahiers Ferdinand de Saussure*, 21, p. 93-96, 1964.

Saussure, Ferdinand de (1894-5?) 3293 =N7), Morphologie. F. de
Saussure: *Cours de Linguistique générale.* Édition critique par
Rudolf Engler, fascicule 4, p. 17-21, Harassowitz, Wiesbaden
1974, English translation, in: Godel (ed.) 1969, p. 26-38.

Saussure, Ferdinand de (1903) Souvenirs de F. de Saussure con-
cernant sa jeunesse et ses études, *Cahiers Ferdinand de Saussure*, 17, p. 12-25, 1960.

Saussure, Ferdinand de (1916) *Cours de Linguistique générale,*
publié par Charles Bally et Albert Sechehaye avec la collaboration de Albert Riedlinger, Payot, Paris 1967.

Saussure, Ferdinand de (1986) *Cours de Linguistique générale,*
édition critique préparée par Tullio De Mauro, Payot, Paris.

Scheerer, Thomas M (1980) *Ferdinand de Saussure.* Erträge der
Forschung, Wissenschaftliche Buchgesellschaft, Darmstadt.

Schuchardt, Hugo (1917) Review of Cours de Linguistique
générale, *Literaturblatt für germanische und romanische
Philologie,* vol. 38, nr. 1-2, Januar-Februar, col. 1-9.

Werlen, Iwar (1981) Hjelmslevs Saussure Rezeption, *Cahiers
Ferdinand de Saussure,* 35, p. 65-86.

Winteler, Jost (1876) *Die Kerenzer Mundart des Kantons Glarus,*
Leipzig, Heidelberg.

"Neu-Romantik" or "Neuro-Mantik"? Psychiatry, Literature, and the Unconscious in the 1880s

Uffe Hansen

Among the various factors which brought about the dissolution of Naturalism in the late eighteen-eighties there is one that is indisputably central: the intrusion of new theories of the human mind which turned the current views on the relation between body and mind upside down. However different forms the reaction against Naturalism might take, there is one feature that is common to Symbolism, Decadence, Impressionism, Neo-Romanticism, and so on: at the core of the new movements you will invariably find a fascination with the almost imperceptible operations of the mind replacing the preoccupation of the Naturalists with the physical world, with the determinants of race, environment and time, the three primordial forces according to Hippolyte Taine. We are clearly dealing with not just a revival of the pre-positivist ideas of the Romantic Movement, "Neu-Romantik" (neo-romanticism), but with a "Neuro-Mantik", an anthropology based on a psychology which applied the methods of positivism, precise and detailed observation, to the functioning of the observer's psyche and especially to the mental processes near or below the threshold of consciousness.

No doubt, in the history of literature this is a commonplace which can be found with few variations in all descriptions of the anti-naturalistic movements. The open question, however, is: What is the exact meaning of the "Kunst der Nerven" (Wunberg 1968: 89), the new psychological naturalism, the alleged new concept of the unconscious? Since "psychology" is the watchword of the young generation (cf. Wunberg 1968: 50) we might expect to find in the surveys dealing with this period references to at least some of the great psychologists of the 1880s, and an indication of the exact contents of their theories. More often than not, however, we have to be content with mere name-dropping: Charcot - who was a physiologist and a neurologist and definitely not a psychologist; William James, who coined the phrase "stream of consciousness", but whose chief work, "The Principles of Psychology", published in 1890, was mainly a comprehensive exposition of the psychological research of the decade; and Bergson, finally, although the influence of his works belongs to a much later period. In a recently published and otherwise fine book, "Die literarische Décadence", by the renowned German literary historian Wolfdietrich Rasch, even these names are missing, and this is rather the rule than the exception. Even the brilliant essay by Marianne Kesting, "Der Abbau der Persönlichkeit" (The dissolution of personality) which deals with the origins of modern drama, only mentions Charcot and Théodule Ribot in a short paragraph (Kesting 1976: 212). Literary criticism apparently suffers from a sort of phobia towards the psychology of the years when psychology was established as a science in its own right, freed from its complete dependence on neurology.

Malcolm Bradbury and James McFarlane, for instance, in their essays on the origins of Modernism limit themselves to emphasizing the role, of intuition and sensitivity in the changing psychological climate of the 1880s, as in the following quotation:

> The dissolution of Naturalism is perhaps best caught in an image. In the lamplight of Maeterlinck's "L'Intérieur "(1894), an old man broods sadly on a girl's suicide by drowning. 'Why?' he wonders. 'Nobody knows. What can anybody know?... You can't look into the soul as you can into a room.

Here, with that phrase, is both the nature of the new challenge and the defining limitations that the writers who sought to press beyond Naturalism into Impressionism, or Symbolism, or Psychologism must have felt: here is the continuity and the rupture.

And the two authors conclude:

> But to this soul - mind, spirit, nerves, psyche, consciousness - there is no fourth wall to be opened to permit direct observation: nothing immediately audible, visible, accessible to the five standard senses. And on the other side, there was much that was beyond the reach of reason (Bradbury/McFarlane 1986: 195-197).

Much the same attitude can be found in current Scandinavian expositions of the change from the Naturalistic eighties to the Symbolistic nineties which stress the Nietzschean influence, beginning with Georg Brandes' lectures on the philosopher in 1888. That is: no new psychology, but rather awe of the mystery of life:

Oh, Mensch, gib Acht,
Was spricht die tiefe Mitternacht?
Ich schlief, ich schlief -,
Aus tiefem Traum bin ich erwacht:-
Die Welt is tief,
Und tiefer als der Tag gedacht

This may be an appropriate expression of how segments of the new generation felt about Naturalism in its crumbling state, but it certainly reveals little about the origins of the need for an alternative psychology in the early eighteen-eighties.

It is important, I think, to distinguish between those tendencies in the eighties which represented a more subtle development of the older physiological Naturalism, and those which meant the substitution of an entirely new model of the human mind. The notions of degeneration and decadence which were favourite ideas of the time dating back to the French physiologist Morel's theory in the eighteen-fifties, the heyday of early Positivism, and Lombroso's contentions about the "born criminal" were similar hereditary theories firmly rooted in orthodox reductionist Positivism. Ideas like these evidently played their role in the last two decades of the nineteenth century, but they were *common* to the Naturalist movement and its opponents.

The same thing can be said about a second constitutive feature of the time: mental self-dissection. In August 1890 the Viennese critic Herman Bahr, one of the foremost spokesmen of the new literary taste, announced in an essay, *Die neue Psychologie*, that

the old psychology only finds the end effect of the emotions, the form that consciousness gives to them and that memory preserves. The new psychology will trace their first elements, the beginnings in the darkness of the soul before they emerge into full daylight, all these long, detailed, intertwined processes of the emotions which at last push their complicated facts beyond the threshold of consciousness in the form of simple effects (Wunberg 1968: 57).

This tracing of the silent operations of the organism is in itself no indication of an altered attitude towards the psychic life of man, nor is the preoccupation with morbid states of mind, insofar as they can be reduced to morbid states of the body. Bahr, although he stressed the need for a new psychology, mentions only one psychologist, Théodule Ribot, and to Ribot the psychic phenomena were ultimately reducible to the elementary physiological processes. Friedrich Fels, another prominent critic, makes this clear, when he says:

> The man who immerses himself in the inner world and traces with anxious care every little nuance of the life of his soul is a Naturalist (Wunberg 1976).

In this respect there is merely a quantitative difference between for instance the physiologist Claude Bernard (1813-1878) with his literary disciple Émile Zola and the self-observing Narcissus-figures in several novels of the late eighteen-eighties like Paul Bourget's. The only difference is the refinement of the tools for the observation of inner processes.

So far, we can take two facts as established: neither the concept of intuition instead of observation nor the scrutinizing of the almost imperceptible half bodily, half mental states can be the main issues when the young literary generation turned against the older psychology. Do we end up with the contention in Zola's "Le roman expérimental" (1880), that the choice is between "l'homme métaphysique" and "l'homme physiologique"? By no means, for the starting point for the new developments could be found in a special branch of *experimental* psychology, the study of hypnosis. By the hypnotic technique a gate could be opened to direct observation of the operations of the human mind, even such operations as are normally subconscious. Furthermore, the experiments could be conducted in a way which met the fundamental demands of Positivism.

Although hypnotic phenomena are well-nigh universal, modern explorations of hypnosis began through the attention attracted to it by Franz Anton Mesmer (1734-1815), a Viennese physician working in France, who promulgated the doctrine that hypnotic states were induced by so-called animal magnetism, a force he believed to emanate from his own body, especially his hands. This pseudo-scientific theory, working with analogies given by the contemporary discoveries in the field of electricity and assuming the existence of a universal fluid pervading the cosmos quickly spread all over the continent and especially established itself in Germany and France. Not until 1838 did Mesmerism burst upon Victorian Britain, which was one of the main reasons why psychology and psychiatry in England developed along other lines than on the continent.

In spite of the fact that the characteristic features of Hypnotism were thoroughly described right from the start by German and French students of magnetism, all efforts to have this theory and its observations acknowledged by the scientific authorities failed, mainly because the underlying doctrine could not be experimentally proved. Nevertheless, Mesmerism was widely spread not only by a host of itinerant performers who made a living out of it, but also by men of high intellectual distinction, above all among the German Romantic poets and natural philosophers,

Between 1820 and 1880 the development of hypnosis exhibits two main features. Firstly, the intrusion of various obscure or controversial ideas into the theory, phrenology (e.g. in James Braid's case) and since 1849, especially in Great Britain, spiritualism, which contributed much to reducing what was left of the scientific reputation of animal magnetism (cf. Oppenheim 1985). Secondly, the works of a few physicians groping for a better understanding of the enigmatic phenomena and trying to exclude from Mesmerism all such elements as were incompatible with the general views of positivism. I shall mention only one of these positivist pioneers, James Braid, who was the first to use the word hypnosis, in this way underlining that we had to do with a normal condition related to normal sleep. Braid was a Manchester physician who had been much impressed by mesmeric demonstrations given in 1841 by the French magnetizer Lafontaine, although he of course could not agree with the mesmeric basic fluid theory. According to Braid the hypnotic state was brought about by monotonous sensory stimulation of the optic centres of the brain, resulting in a sort of local cortical paralysis which hereupon would cause a general paralysis of all the higher brain functions. Consequently the influence of the "hypnotist" was regarded as insignificant, his only function being to help the patient to concentrate all his attention on the one idea of sleep, thus facilitating auto-hypnosis. Any notion of a kind of mental or hidden physical influence on the part of the hypnotist was dismissed beforehand (cf. Bramwell 1897).

Although Braid's theory as he formulated it in his book from 1846, *The Power of the Mind over the Body, an experimental Inquiry into the Nature and Cause of the Phenomena, attributed by Baron Reichenbach and others to a "New Imponderable"*, was based on brain physiology and therefore acceptable to medical circles, his research in this field was soon practically forgotten. Even such promising aspects of hypnosis as the complete anesthesia which could be brought about and which was actually successfully applied in surgery, became obsolete soon after when ether and chloroform anesthesia were discovered.

Braid died in 1860 and for 20 years even his name remained unknown among neurologists and psychiatrists. In the late eighteen-seventies, however, a certain recognition of Mesmerism was slowly making its way in connection with the study of hysteria. In 1878 the French neurologist Jean-Martin Charcot stated that the hysterical phenomena of lethargy, catalepsy, certain types of hallucinations, and transient changings of personality could be provoked by hypnosis, asserting that the very ability to be hypnotized was in itself a genuine neurosis. Charcot's description of the cause and the effects of the hypnotic states was given in purely neurological terms, as were his demonstrations that hysterical conditions under favourable circumstances could be cured by hypnosis. What he did not realize or did not want to face was the extent to which the contents of the hypnotic state were determined by the hypnotist, or even had to be learned by the hypnotized subject. In short, the psychological elements in the whole setting. The important thing about Charcot's studies of hysteria was that his immense authority as a neurologist made it possible for other physicians to recognize hypnotism and use this old technique for therapeutic and experimental purposes without being considered quacks. But not Charcot, although his investigations aroused enormous interest, brought the proper revival of hypnotism and an adequate understanding of the fundamentally psychological nature of it.

In 1879 a Danish stage hypnotist, Carl Hansen, went from town to town in Germany organizing theatrical hypnotic performances which attracted large crowds. In spite of the fact that Hansen on several occasions willingly put himself and his subjects at the disposal of interested physicians and scientists, the academic world generally looked on his hypnotic demonstrations as a product of fraud, a conjurer's skill, or the credulity of simple-minded participants and onlookers.

However, in Breslau the distinguished neurologist Rudolf Heidenhain was impressed by Hansen's results and adopted his method for his own investigations (Bernheim 1888: 110). The Danish hypnotist seems to have been no mere average mesmerist performer. In hypnotizing his subjects he employed the method of letting them gaze at a small luminous object, i.e. the method applied by Braid, which is a rather remarkable feature, Braid and his theories having been almost forgotten by that time. But Hansen contended, contrary to Braid, that this was not sufficient to bring about the full range of the phenomena of artificial somnambulism, that, furthermore, hypnotism had nothing to do with the fluidist assumptions of magnetism, and finally that the person of the hypnotist played an important role in the process, which Braid had definitely denied. In short, that a sort of psychic influence, a powerful interpersonal agent was indispensable.

It is interesting to notice what Heidenhain, and about the same time his colleagues in Chemnitz, could use from Hansen's performances and what they rejected as a priori unscientific (Heidenhain 1880), because here we can observe two main aspects of positivism. Hypnosis was adopted as a technique of research because it would make controlled experiments possible, thus satisfying the demands of positivist method. But apart from being a scientific method positivism, although this strictly speaking could not be deduced from its professedly anti-metaphysical position, generally included an ontology as well, which recognized only physical phenomena as real and considered all psychic phenomena as ultimately reducible to basic physical elements.

The conviction that they had to anchor themselves firmly in physical facts forced the German students of hypnotism at this time to restrict themselves to examining the purely physiological aspects of the abnormal condition and to keep all psychological considerations on the role of the hypnotist within the confines of natural science.

Although, consequently, the psychological core of hypnosis was passed over in silence even by those who did accept the reality of the observed phenomena, not all onlookers were completely satisfied with the current scientific explanations. In Vienna, where Carl Hansen gave a number of hypnotic performances in February, 1880 - three weeks later they were forbidden by the police - Sigmund Freud, then a student of medicine, was much impressed by the experiments he had witnessed, and he began adopting hypnosis as a therapeutic technique five years later (Jones 1953: 258). A few months after Hansen's departure from Vienna, the physician Josef Breuer began treating what has become the classic case of hysteria, that of Fräulein Anna O.. In his psychotherapy using hypnosis, Breuer was able to recognize the presence of two distinct alternating states of consciousness in his patient: one a fairly normal one, the other that of a troublesome and naughty child. What to us, a hundred years later, might seem to be the first hint of a basic concept in psychoanalysis, the overwhelming importance of early childhood experience, was in fact already one of the controversial issues of hypnotism, the phenomenon of age regression, a consequence of the increased capacity of the memory in hypnotized persons, the so-called hypermnesia. However, in spite of these two prominent exceptions the Viennese climate was not at all favourable to Hansen's ideas.

During his campaign he also visited Strassburg and Nancy, where his performances were witnessed among others by the professor of internal medicine at the University of Nancy, Hippolyte Bernheim (Bernheim 1888: 110). Shortly afterwards Bernheim heard about a country doctor, Auguste Ambroise Liébeault, who had then for years successfully treated hundreds of patients with hypnosis, not the Braidian auto-hypnotic technique but mere verbal suggestion, inducing the hypnotic state by concentrating the attention of the patients on the one idea of falling asleep. The application of suggestion, verbalized or not, the purely mental influence on the patients' mind, marks the starting point of a new paradigm, a new model of human personality. It is in itself remarkable that a renowned professor should adopt a heretofore ill-reputed method going all the way back to the era of the speculative medicine of Romanticism from an old obscure doctor and perhaps ultimately from a stage hypnotist. I think that neither Carl Hansen nor Liébeault should be over-estimated. Probably they were mere catalysts in a process that must have taken place over a long period. Either the experimental method of positivism had to make the human mind its object, or the positivist reductionist model had lost its capacity for integrating the sum of observations which had accumulated in the course of a century since the first groping mesmerist experiments. Of course this is a mere hypothesis; another intriguing explanation would be that the human mind during the nineteenth century had undergone such changes in its very structure that new, or rather simply neglected phenomena forced their way into scientific psychology.

After having written some short reports on the subject, Bernheim published his principal work in 1886, *De la suggestion et de ses applications à la thérapeutique*. A German edition appeared in Vienna in 1888 with a 10-page preface by the translator, Sigmund Freud. In the author's preface Bernheim contended, against Charcot and his followers at the Salpêtrière hospital, that all the alleged physiological causes and effects of hypnosis after careful examination would turn out to be the manifestations of the purely mental influence exerted by the hypnotist. Further, he clearly stated that the capacity of being hypnotized was not just a pathological condition specific to hysterics but a constituent part of normal human personality, the broad field of suggestibility. Thereby he defined the concept of suggestion as the capacity of one mind to arouse ideas and concomitant perceptions, actions, and physiological alterations in another person by merely verbal means, and that suggestion differed from other sorts of psychological influence - like ordering, informing or teaching - in that the ideas called forth in another brain would not be subject to inquiry concerning their origins, but would be accepted, as if they had spontaneously arisen out of the brain of the hypnotized person. By stressing that suggestibility was an important part of the normal mind, the school of Nancy and its adherents in France, Switzerland and Germany claimed that the hypnotic technique could be employed as the basic approach to the unconscious mind and consequently to a new understanding of human personality altogether.

I shall not at this place give any broad survey of Bernheim's observations and theories (cf. Barrucand 1967) but confine myself to pointing out the most important features marking the replacement of the older model of personality by a new one. Firstly, the observation of the peculiar psychological relationship between the hypnotist and the hypnotized, the so-called rapport. In the hypnotic state the hypnotized person, whether apparently asleep or awake, would be oblivious of all but the hypnotist and could perceive the outside world only within the limits set by him. This 'rapport' extended its influence beyond the session, above all in the post-hypnotic suggestions. That is, a hypnotized subject would, in most cases, in his wakeful state execute an order given to him during his hypnotized state. The induced idea would mostly, but not always, remain unconscious until the order was being executed. However strange or even absurd these actions might be, the subject invariably would successfully rationalize what he had done, thus seemingly integrating the isolated action into his normal conscious life. Secondly, the fragmentation of memory, the different forms of amnesia, which separated normal conscious life from the hypnotic states. The person who is put into hypnosis remembers all that happened during the previous sessions, whereas the waking self retains no memory of these periods. The hypnotist thus calls forth in his subject a special life of its own, separate from the normal conscious life.

The new model resulting from these observations was based on the conclusive evidence of the fundamental duality of conscious and unconscious mental strata, the existence of unconscious mental operations having been experimentally proven, and not just hypothetically assumed as in the natural philosophy of Romanticism (cf. Freud 1916: 118-120). If, moreover, the unity and coherence of human personality, as it was generally assumed, depended on the unity and coherence of memory and on comparatively stable character traits, the emergence of different distinct personalities, each with its own memory, during the hypnotic sessions (or on rare occasions spontaneously) would indicate that "personality" in fact was more like a cluster of personalities, consisting of several subliminal and one manifest personality (cf. Ellenberger 1970: 126-147). Frederic Myers, a prominent member of the British "Society for Psychical Research" even went so far as to declare:

> Our habitual or empirical consciousness may consist of a mere selection from a multitude of thoughts and sensations, of which some at least are equally conscious with those that we empirically know. I accord no primacy to my ordinary waking self, except that among my potential selves this one has shown itself the fittest to meet the needs of common life (Oppenheim 1985: 257).

A further important aspect was the understanding of mental phenomena as being irreducible to biological or physical elements, as they obeyed other laws than purely physical ones. The assumed unity of body and mind had been clearly at work when the organism-model of personality was established at the end of the eighteenth century. Analogous to the self-evident slow and continuous evolution of the physical organism out of the inherent predispositions, the idea of a correspondingly cohesive unfolding of the character of an individual was generally accepted. The positivist axiom of the psycho-physical parallelism was completely in accordance with this paradigm. However, by the mid-eighteen-eighties the experiments with hypnosis seemed to invalidate not only the concept that mental processes could be explained in physiological terms, but also the traditional criteria for defining personality as a dynamic unity. This was an open challenge to the prevailing materialistic monism and to the still persisting belief in the unity and autonomy of the Cartesian *res cogitans* (cf. Hansen 1986: 47-108).

Of course we cannot expect that the majority of psychiatrists, philosophers and poets in the eighteen-eighties were ready to abandon basic convictions that had been deeply rooted in European culture for such a long time. In fact, the prestige of natural science and its methods, *and* of the materialistic ontology were hardly troubled by the peripheral investigations and seemingly wild speculations of a few extravagant psychiatrists and philosophers. On the other hand, the mass media abounded in articles on the exciting new topic, not least in reports of spectacular criminal cases in which the delinquent allegedly had acted under hypnosis or in a state of more or less spontaneous split personality. On a somewhat higher literary level, Robert Louis Stevenson's *Dr. Jekyll and Mr. Hyde* (1886) is just one instance among hundreds. (Incidentally, it should be noted that the ill reputed Mr. Hyde cannot simply be dismissed as the wicked or instinctual impulsive part of the high-principled doctor. He certainly is not very admirable, judged by moral standards; on the other hand, and this point is stressed by the author several times, he is indisputably not only younger and more vital than his alter ego, but even superior to him in regard to versatility and intelligence. In this respect Mr. Hyde corresponds more to Frederic Myers' subliminal self, i.e. a latent but not necessarily inferior personality, than to an instinct-ridden neurotic of Freudian provenance.)

This public curiosity and the host of misunderstandings and exaggerations contributed strongly to bringing the new ideas concerning human personality into disrepute among members of the learned professions. History repeated itself: precisely the same thing had happened to the investigators of animal magnetism in the course of the eighteen-twenties and -thirties, when well-established psychological facts and psychical research were constantly confused.

Nevertheless, some of the best brains of the decade adopted the new psychology or at least some aspects of it. The impact of the new concept of personality can for instance be recognized in the aphorisms of Nietzsche from the mid-eighties, in which the older Nitzschean opposition (as in *Die Geburt der Tragödie*) between the formless chaotic Dionysiac principle and the Apollonean principle of individuation is now further complicated by his fascination with the idea of the human soul as a "multiplicity of subjects" (Nietzsche 1966 Vol. III: 473). Nietzsche's dependence on the psychological paradigm that came into being about 1880 has hardly been recognized, so that literary historians generally consider him as the most important founder of the new concept of man (cf. Kesting 1976). The main reason for this may be that he only reluctantly, if ever, admitted his intellectual indebtedness to anyone.

In contrast to him, another of the partisans of this controversial psychology, the Swedish dramatist August Strindberg, time after time frankly pointed to Hippolyte Bernheim (and to a lesser degree Théodule Ribot) as his psychiatric authority. Especially in his so-called naturalistic dramas between 1887 and 1890, the direct influence of the School of Nancy is omnipresent. In spite of the general admiration for these milestones in the history of drama, critics have not ceased to point out the multitude of inconsistencies which mar the plots and the development of the characters. These alleged deficiencies, however, show exactly where Strindberg stands: neither as an adherent of Hippolyte Taine's or Théodule Ribot's physiological psychology, which was what Georg Brandes wrongly assumed, nor as a creator of whole and consistent characters along the lines of, say, Sainte Beuve.

Central to Strindberg's idea of the nature of human personality is the conviction that a person's mind is infinitely permeable, that a flow of suggestions and mental energy passes from one mind to another. This mental osmosis, as you might call it, decomposes the characters of his plays to the extent that, for instance, the Captain in *The Father* is not simply one developing personality, but a disconnected series of sub-personalities, rapidly replacing each other and bringing to the surface forgotten, abandoned or latent strata of his psyche. What makes the play so uncanny is above all Laura's great skill in provoking these radical changes. However, this skill or power of will is by no means a definite and stable personal faculty of hers, but the result of her gradual absorption of her husband's strength and faculties.

Without taking the frame of the new psychological paradigm into consideration one can hardly adequately understand Strindberg's characters. Judged by the standards of common-sense psychology the Captain is nothing but a lunatic, possessed by antifeminist delusions, and Laura, correspondingly, a monster created by a misogynist's imagination - which still seems to be the general opinion among critics.

In his second naturalist drama, *Miss Julie*, Strindberg declared in his preface: "My characters are conglomerations of past and present stages of civilization, bits from books and newspapers, rags and tatters of fine clothing, patched together, as is the human soul." This could be, and has been, misunderstood as if Strindberg only intended to stress the elusive and indeterminable character of the mind, as did for instance Knut Hamsun in a famous essay a couple of years later (*From the Unconscious Life of the Mind*), describing the "untrodden, trackless journeying by brain and heart, strange workings of the nerves, the whisper of the blood, the entreaty of the bone, all the unconscious life of the mind." What the Norwegian spokesman of the young literary generation propagates is, all things considered, the physiological psychology of the positivists Fechner, Helmholtz, Taine and Ribot, in short, the position of the "old" naturalism.

The psychological problem in *Miss Julie* is of a different nature. Julie does not live her "own" life, her so-called character being put together of conflicting sub-personalities. Most important among these are the scraps of her father's and her mother's minds, because these mental formations are constantly endeavouring to annihilate each other. Instead of being reduced to a mental battlefield for her parents' conflicts, Julie desperately tries to live a life of her own, so to speak, but is incapable of bringing any coherence into the rapidly changing series of incompatible attitudes. She is permanently assimilating elements from the minds of her partners. In addition to this self-destructive process another conflict arises, when her elusive, indeterminate mental life clashes with the strict determinism prevailing in the physical world, to which her body belongs.

At the end of the play she realizes: "I just have no self. I don't possess a thought which I haven't got from my father, no passion which I didn't get from my mother." In the traditional sense of the word there are no distinct "characters" in this work, but merely a constellation of physical individuals, a crowd in which ideas, emotions, and volitions are circulating. - It is no coincidence that Bernheim's psychology of suggestion leads to a study of the psychology of the crowd a few years later, Gustave Le Bon's *"Psychologie des foules* (1895). - Compared with this dominating motif, the much debated theme of the war of the sexes moves on the surface, being a simple consequence of the far more comprehensive problem of identity.

Strindberg characterized the six plays between 1887 and 1890 as naturalistic, thereby indicating that to him their transpersonal psychology stood on the firm ground of the scientifically conducted experiments of the Nancy school. However, in his later works he took the further step of radicalizing his idea of the multiplex and mutable nature of that ephemeral entity which is usually called the personality of man. His pan-psychism, especially after the turn of the century, now tended to incorporate metaphysical ideas of an all pervading psychic substance. This development clearly indicates what was going to put an end to the scientific study of the human mind along the lines of the school of Nancy: the neutralization of the new paradigm by linking it with occultism. But in the 1880's the psychology from which Strindberg got his inspiration, was still using the positivist method of observation and verification, and, like other sciences, repudiated anything metaphysical or supra-natural.

McFarlane, as one instance among many others, underestimates the controversial nature of the results of the Nancy psychologists when he fails to distinguish between Bernheim and the orthodox reductionist psychology of the decade. That the gap was felt very acutely at that time appears from the controversy at the first international congress on hypnotism, which took place in 1889 in Paris. Pierre Janet, who adhered to Charcot's positivist views, condemned Bernheim's assertions of the purely mental character of hypnosis, because, in his opinion, they entailed the elimination of any kind of determinism and he declared them anti-psychological, because psychology, like physiology, also had its laws. Bernheim replied to this that there was one basic law: that any brain cell activated by an idea tends to bring the ideas into being.

This psychology, positivist in its method of investigation, but non-reductionist in its conclusions, was the point of departure for modernism. Not the universal struggle between a scientific and an intuitive "Weltanschauung", although this dichotomy lingers on, even in recent essays on the culture of the "fin-de-siècle". That is: not a revival of Romanticism, "Neu-Romantik", but a genuine "Neuro-Mantik".

Bibliography

Barrucand, Dominique (1967) *Histoire de l'hypnose en France*, Bibliothèque de Psychiatrie, Paris.

Bernheim, Hippolyte (1888) *Die Suggestion und ihre Heilwirkung*, übersetzt von Sigmund Freud, Leipzig und Wien. Reprinted 1985, Tübingen.

Bradbury, Malcolm and McFarlane, James (eds.) (1986, 1976), *Modernism 1890-1930*, Pelican Guides to European Literature, Harmondsworth.

Bramwell, J. Milne (1897) James Braid; his Works and Writings. *Proceedings of the Society for Psychical Research*, Vol. XII, pp. 127-166, London.

Ellenberger, Henri F. (1970) *The Discovery of the Unconscious. The History and Evolution of Dynamic Psychiatry*, New York.

Freud, Sigmund (1916) *Vorlesungen zur Einführung in die Psychoanalyse* Freud-Studienausgabe, Band I, Frankfurt am Main, 1962.

Hansen, Uffe (1986) *Conrad Ferdinand Meyer: "Angela Borgia". Zwischen Salpêtrière und Berggasse*, Bern.

Heidenhain, Rudolf (1880) Der sogenannte thierische Magnetismus. *Wiener medizinische Blätter,* pp. 169-172, 199-202, 222-223, Wien.

Jones, Ernest (1953) *Sigmund Freud: Life and Work*, London.

Kesting, Marianne (1976) Der Abbau der Persönlichkeit. Zur Theorie der Figur im modernen Drama. *Beiträge zur Poetik des Dramas* herausgegeben von Werner Keller, Darmstadt.

Nietzsche, Friedrich (1966) *Werke in drei Bänden* herausgegeben von Karl Schlechta, München.

Oppenheim, Janet (1985) *The other World. Spiritualism and psychical Research in England, 1850-1914*, Cambridge University Press, Cambridge, London, etc.

Ribot, Théodule (1884) *Les maladies de la personnalité*, Paris.

Wunberg, Gotthart (1968) (ed.) Hermann Bahr. Zur Überwindung des Naturalismus. Theoretische Schriften 1887-1904 *Sprache und Literatur* 46, Stuttgart und Berlin.

Wunberg, Gotthart (1976) (ed.) *Das Junge Wien. Österreichische Literatur- und Kunstkritik 1887-1902*, Tübingen.

Monument of Empire: Aesthetics, Politics and Social Engineering in the Writings of John Buchan

Preben Kaarsholm

In spite of its never attaining to total control or influence in a society, imperialism in its programmatic form, as it develops in Great Britain in particular in the years just before and around the turn of the century, shares many characteristics with and can be seen as a forerunner of later European totalitarian ideological and political systems.[1]

One instance of this is the politicisation of aesthetics which occurs at an emphatic scale in fascist and Stalinist societies in the 1930s and 1940s, but which can also be observed to be in operation in the culmination period of formal or self-proclaimed imperialism decades earlier.

In Great Britain, the most powerful of the imperial states in this period, and the one in which imperialism as a movement reaches its highest level of both formulation and impact, it is remarkable that, far from being only a movement of national chauvinist optimism, imperialism becomes also a movement for cultural crisis management and emerges out of widespread *fin-de-siècle* experiences of decadence and cultural pessimism.

In this context, the imperialist politicisation of aesthetics evolves as the creation of symbolical representations or allegories of the preconditions for a revitalisation and reordering of culture and society which often assumes 'monumental' forms.

Monumentalism can be found in the music of the period, e. g. in the works of Edward Elgar from his "Imperial March", written in 1897 to celebrate the Diamond Jubilee of Queen Victoria, through his 1901-07 "Pomp and Circumstance Marches" to ceremonial pieces like "The Crown of India Suite" and the

---

[1] For a discussion of similarities and differentiations between imperialism and totalitarian programmes cf. Arendt 1973, Part Two, and the introduction to Kaarsholm 1987a.

"Coronation March", both from 1911 (cf. Moore 1984a: 145ff.; Moore 1984b).

It comes out in architecture in a whole set of examples both in Britain and in the colonies - prominently in Edwin Lutyens' and Herbert Baker's majestic buildings of state in India and South Africa, but also in less obvious instances as Lutyens' experiments with different "vernacular" styles in his projects for private houses, which express the endeavour of a return to the organic vitality of the English past (cf. Hayward Gallery 1981; Irving 1981).

In literature, monumentalist politicisation appears in the bombastic counter-decadent poetry of William Ernest Henley (cf. Buckley 1945; Guillaume 1972), in Kipling's solemn celebrations of the imperial ethic in poems like "Recessional" (which contrast strangely with his statement in a letter to Cecil Rhodes in 1901 that "England is a stuffy little place mentally morally and physically") and in his numerous stories and tales from 1890 onwards, in which he represents the organic imperialist fusioning of sound European and 'barbarian' energies as opposed to the centrifugal and catastrophic tendencies of decadent liberalist idealism on the one hand and fanatic oriental savagery on the other (cf. Kaarsholm 1987b).

Similar fears and hopes for an authoritarian reconstruction can be found in Joseph Conrad's "The Nigger of the Narcissus" from 1897, where society - like a ship caught in a storm, calls for strong, hierarchical structures of command to save it.[1]

Finally, to mention just one popular cultural *genre*, the journalism of George Steevens provides a running argument of a similar nature for the return to a healthy brutalism by means of imperialism and perhaps better than anything else illustrates the new monumentalism and politicisation of aesthetics. Steevens' report for the *Daily Mail* of the 1897 Diamond Jubilee is a prime example of this (cf. Kaarsholm 1985).

At the same time we also find in the programmatic imperialism of the turn-of-the-century period the reverse movement - of politics becoming infused with and structured by aesthetics.

This tendency can be observed in a whole range of attempts to promote a new national and imperial politics which is governed, not by 'traditional' class and party contradictions, but by new

---

[1] For the ambiguities of Conrad's treatment of the theme of imperialism cf. Parry 1983.

relations between elites and masses that are at one and the same time 'modern' and 'conservative'.[1]

Promotion of the new politics takes place, characteristically, in what was called at the time the "New Journalism" of papers such as the *Daily Mail* and in the monumental and ceremonial organisation of political spectacles like those set on foot to celebrate Queen Victoria's Jubilees or the Coronation of Edward VII in 1902. But more importantly, perhaps, the aestheticisation of politics asserts itself at the level of the contents of debates on political and social reform matters. Here increasingly discussions concern themselves, not so much with the purposes and rationalities of programmes and arguments, but rather with their form, with the ethos and mechanism of 'efficient administration' in itself, and where simultaneously democracy and egalitarianism come to be seen as hindrances to progress and 'efficiency'.

This tendency finds exemplary theoretical expression in a book brought out in 1901 by Arnold White, *Efficiency and Empire*, and comes to the fore more pragmatically in 1902-3 in the efforts of the "Coefficients'" group, which brings together members of parties and political organisations that had previously been separated by the boundaries of the 'old' politics (cf. Searle 1971; Semmel 1960).

In practical and administrative terms a similar trend can be observed in the attempts by Alfred Milner to set up an ideally ordered construction of efficient colonial administration in South Africa in the years following the Anglo-Boer War and in the imperialists' almost cultic idolisation of Milner in Britain in the years around 1900 (cf. Marks and Trapido 1979; Schröder 1978).

Milner's attempts to create a monument to social and administrative efficiency in South Africa, to make of South Africa a 'model' colony and system of government represents one of the most direct forms in which the legacy of imperialist thinking is still with us today: Milner's reforms made an important contribution to the political artefact that was established in the Union of South Africa in 1910, and which, in its turn, provided the foundations for the modern distribution of wealth and power among groups and races in South Africa and, in its authoritarianism and ideology of 'separate development', for the modern apartheid state.

---

[1] For similar constructions of 'reactionary modernism' in fascist Germany cf. Herf 1984.

The following discussion focusses more closely on one particular example of the fusioning of aesthetics and politics in the period of programmatic imperialism and examines aspects of 'monumentalism' in the writings and career of John Buchan.

Buchan is an interesting figure as regards the interaction between aesthetics, politics and social engineering, because he was so closely linked to the British political establishment and in the most direct way wrote to function as an ideologue, whose books and articles were intended to influence and form the political attitudes and moral outlook of their readers.

Buchan's development as a writer was intimately linked to his career as politician and civil servant in a variety of ways. It was his achievement as a journalist and writer of fiction that provided the basis for his being accepted into the social life of the British governing elite. His 'theory', as formulated in his novels, short stories and essays, was closely related to his 'practice' as an administrator - from his work with Milner in South Africa between 1901 and 1903 to his initiatives as Director of the War Cabinet's Department of Information in 1917.

This close interrelationship between 'theory' and 'practice' also asserts itself in the form and *genre* of certain of his writings: *A Lodge in the Wilderness* (1906), for example, is a mixture of fiction and non-fiction, an imperialist political treatise given the form of a Platonic symposium held by figures that to Buchan's mind represented the quintessential characteristics of the British social elite.

The psychological element perhaps should not be exaggerated in the interpretation of Buchan's life and works. On the other hand, prominent features of his career development as well as important themes and contradictions in the texts he produced do hardly make sense unless seen against the background of the specificities of Buchan's personality. Furthermore, this personality can be argued to be of interest, not so much as an individual psychology, but as a 'social character' which was formed in a particular time and place and determined by a particular history.

In any case, Buchan's career, his peculiarly close and uncritical relationship with the upper strata of the social hierarchy and with the nature of that hierarchy in itself, can hardly be understood without realising how his whole life became a 'monument' which was structured around a process of social climbing that took him from a relatively humble Scots Calvinist background through Oxford and journalism into the governing elite of higher civil servants, politicians, club members and aristocrats.

As it comes out in both his published writings and his private letters, Buchan's snobbism probably strikes most people today as unbearable. The snobbery, however, expands into a broader romantic pessimism that turns Buchan into a critic of the new middle classes whose social status is only founded on money and makes him yearn for an older, more authentic hierarchy, based on merit and respect. This order of society he refers to as that "whole lost world of pastoral" which he thought existed in the countryside of the Borders where he grew up - a 'natural' feudalism and cohesion in society which he found was being gradually replaced by "the vulgarity and the worship of wealth." Already at his return from South Africa in 1903 he "began to have an ugly fear that the Empire might decay at the heart" (Buchan1940: 24, 94 and 128).

In spite of his fascination with the social hierarchy and his efforts to serve the civilising mission of imperialism, John Buchan thus repeatedly turned towards *nature* and a simple organic life as antidotes to over-civilisation and the corrupting influences of modern society. Going for walks on the Scottish moors and on the highveld in Transvaal and mountaineering in the Alps he searched for, what in *The African Colony* he described as "the type of pure world before our sad mortality had laid its spell upon it" (Buchan 1903: 103).

Buchan's imperialism became a programme for the reform of society which would bring about a more 'natural' balancing of civilisation and 'barbarism' and thereby act against the crisis of culture that Buchan saw as threatening society. The following presentation of some of the central themes and contradictions in his literary production, which comprises more than a hundred books and pamphlets as well as over three hundred articles and contributions to books, will concentrate on *The African Colony* (1903), *A Lodge in the Wilderness* (1906), *Prester John* (1910) and the most important of the Richard Hannay novels - *The 39 Steps, Greenmantle* and *Mr. Standfast*, which were written during World War One.

Achievement: John Buchan with his mother at the climax of his career as Governor-General of Canada-Quebec in Ottawa 1936.

The thin line that divides "civilisation" from "barbarism": John Buchan as Chief Eagle Face of the Blood Indians in 1937.

The Hannay stories are important in British literary and cultural history as representing a popular literature written for a mass audience by an elite author, representing a mediating instance through which establishment attitudes were given broader effect. The contrast between the sales figures of Buchan's books written for his own social group and of those written for a popular audience is striking: Whereas e. g. *The African Colony* sold just over 500 copies within two years of its publication in 1903 and then 'died', the Hannay novels have been sold in millions of copies and are available to this day.

The common essential theme of the Hannay series may be described as 'the thinness of civilisation'. Their basic thematic contradiction lies between (British) 'civilisation' and 'barbarism' or 'dissolution', but the main point seems to be not so much the triumphant victory of 'civilisation' as its fragility: The thrill of the novels resides in their showing how closely related their protagonists of 'civilisation' are to the forces of 'dissolution' - the vertigo produced by getting to the brink of the abyss of engulfment.

The structural pattern of the plots is classic: 'Order' is threatened by 'chaos'; 'chaos' is fought back; 'order' is reestablished in a new and more elaborate form. It is important, then, to see how 'order' or 'civilisation' and 'chaos' or 'barbarism'/'dissolution' are defined as themes in the novels.

The order of 'civilisation' in Buchan tends to be synonymous with a British upper-class existence: Club life in London, weekends at big country houses, intact networks of elite cooperation, an intricate set of rituals and codes of behaviour and the keeping up of *facades* and appearances as exemplified by the elaborate processes of dressing and describing their outfits the protagonists go through.

The implication seems to be that an intact and confident elite signifies a healthy and well-established social hierarchy. 'Order' in the form of a strong and secure power structure is reinforced and vitalised through imperialism and the Empire.

Moving beyond nationalism and imperialism, 'order' and 'civilisation' take the form of a male fraternity, represented by Hannay and his group of fellow activists. What is being emphasised here is partly the masculine nature of 'order', but also its basis in the organic interaction between different layers in the hierarchy - from the rustic Boer Pienaar and the pragmatic American Blenkiron to the British aristocrat Arbuthnot.

Finally, 'civilisation' is identified with the keeping up of formalities. A recurrent motive is the importance of dress and costume, where the crisis of 'order' shows itself in Hannay's missing his normal clothes, and its reestablishment in his being able to dress correctly again.

The opposing theme of 'barbarism' or 'dissolution' is introduced in the Hannay novels through a group of different motives. One is 'the German threat' as represented by the brutal and homosexual Stumm in *Greenmantle*, but Buchan's attitude towards Germany and Prussianism is always ambiguous, and 'bad' Germans like Stumm are counterbalanced by 'good' specimens like Gaudian in the same novel.

Secondly, the forces of 'dissolution' are represented by 'the Orient', which asserts itself e. g. in the Moslem sects that Hannay and Sandy Arbuthnot are up against in Istanbul in *Greenmantle* or in the form of Jewish influences and conspiracies.

The most spectacular instance of the latter is the plot that Scudder reveals to Hannay in *The 39 Steps*, where Buchan comes close to classic anti-semitic mythology:

> The Jew is everywhere, but you have to go far down the backstairs to find him. Take any big Teutonic business concern. If you have dealings with it the first man you meet is Prince *von und zu* Something, an elegant young man who talks Eton-and-Harrow English. But he cuts no ice. If your business is big, you get behind him and find a prognathous Westphalian with a retreating brow and the manners of a hog. He is the German business man that gives your English papers the shakes. But if you're on the biggest kind of job and are bound to get the real boss, ten to one you are brought up against a little white-faced Jew in a bathchair with an eye like a rattlesnake. Yes, sir, he is the man who is ruling the world just now... (Buchan 1915: 11-12).

Thirdly, 'dissolution' takes the form of independent and active womanhood, as exemplified most prominently by the evil Hilda von Einem in *Greenmantle*, who is almost too much for our fraternity of heroes. Hilda's counterpart, the 'good' woman, is Mary in *Mr. Standfast*, but Mary's 'goodness' stems mainly from her being so very much "like a gallant boy": It is this boyishness, together with distinguished family background that makes Hannay fall in love with her.

Hilda von Einem is the real woman, whose evil influence is so powerful that Blenkiron is unable to withstand it and Richard Hannay only through mobilising his utmost powers of self-control:

> ... again her strange potent eyes fell on my face. They were like a burning searchlight which showed up every cranny and crack of the soul. I felt it was going to be horribly difficult to act a part under that compelling gaze. She could not mesmerize me, but she could strip me of my fancy dress and set me naked in the masquerade (Buchan 1916: 175).

Behind this scare of women lurks a more generalized fear of the ego dissolving or being engulfed in passion, which surfaces in the recurrent episodes in the novels where the protagonists are losing themselves in hypnotic trances or involuntary intoxication.

Through metaphorical disguises such as these, sexuality can be argued to represent the fourth motive in which the theme of 'dissolution' is brought to attention. A fifth 'dissolution'-motive is that of the social rebellion which directly challenges the hierarchy and the organic nature of 'civilisation' - a motive that dominates in *Prester John*.

A sixth motive belonging in the same thematic context is the threat of losing one's position in society - of falling down into the proletariat. As was mentioned, dress and costume play important parts in the Hannay novels, not least in the form of the many disguises Richard Hannay assumes on his secret missions. These are sometimes so effective that they make him doubtful about his true identity. In *Mr. Standfast*, for example, when Hannay is on his way back to London from Scotland, travelling third class on a very full train, his disguise as drunken bagman suddenly provokes desperate fears in him:

> I had got down too far below the surface, and had the naked feeling you get in a dream when you think you have gone to the theatre in your nightgown. I had had three names in two days, and as many characters. I felt as if I had no home or position anywhere, and was only a stray dog with everybody's

hand and foot against me. It was an ugly sensation... (Buchan 1919: 140).[1]

The protagonist's sense of his social identity is so frail that it is under a constant, latent threat of disintegration - only a very thin line divides 'civilisation' from 'dissolution'. In psychological terms, the major fear in the texts is that of disappearing, of losing identity and status through sexual passion, intoxication, loss of control or a collapse of the social hierarchy.

As pointed out above, theory and practice were closely related in the life of John Buchan. His philosophical outlook was influenced by the Hegel-inspired idealism that was brought forward in the late 19th century by such thinkers as T. H. Green, F. H. Bradley, H. H. Joachim and Bernard Bosanquet, and he admired personalities whose high idealist purpose became manifest not only in words and phrases, but in their actions.

Cecil Rhodes, for example, is presented in *The African Colony* as a kind of ideal idealist - one of the "cyclopean architects of new lands" and a matter of fact man-of-business characterized by

a hungry fidelity to ideals for which we look in vain among the doctrinaires... What filled his imagination was the thought of new nations of our blood living a free and wholesome life and turning the wilderness into a habitable place. He strove not for profit but for citizens, for a breathing-space, a playground for the future (Buchan 1903: 393).

In 1901 Buchan was given the chance to contribute practically to the realisation of his imperial ideals. On the merit of his political journalism he was offered a post in Alfred Milner's South African administration as private secretary, in charge first of the concentration camps, later of land resettlement.

Milner had been made British High Commissioner in South Africa and Governor of Cape Colony in 1897 and was one of the main initiators of the intransigent British policy that brought about the Anglo-Boer War. He stood out as an ideal to imperialistically minded young people of his age and was one of the most prominent representatives of a programme of social imperialism, national efficiency and a strong state to offset the decadence threatening modern society.

---

[1] Sandy Arbuthnot undergoes a similar experience of being 'swallowed-up' in his impersonation of the ayatollah-like Moslem priest in *Greenmantle*.

His work on the after-war reconstruction of South Africa was seen as one of the great imperial experiments of the time - an attempt to build, not only new colonies, but a model society based on imperial political notions. Among Milner's most important goals were 1) the securing of South Africa as a predominantly British settlement through the encouragement of immigration, 2) the reconciliation on the basis of British sovereignty of 'the white races' of South Africa - of Britons and Boers, 3) the establishment of a strong state apparatus that would secure modernisation and economic development based on the gold mines and on capitalist agriculture and 4) an orderly regulation of the relationship between white and black races in South Africa with a view to social stability and a rich supply of cheap labour power (Marks and Trapido 1979; Pakenham 1979; Pyrah 1955: 182-221; Thompson 1960: 4-17).

John Buchan's experiences in South Africa and his views on the reconstruction were described in a series of articles and more fully in his book *The African Colony* (1903). It is worth trying to compare the world-view and main ideological themes of this book with those formulated in Buchan's fictional writings.

The African Colony is divided into three parts - "The Earlier Masters", "Notes of Travel" and "The Political Problem". As in the Hannay novels, the basic thematic conflict is the contradiction between 'civilisation' and 'dissolution', and the whole argument of the book can be said to deal with the necessity and practical possibilities of strengthening the thin line that divides the two poles.

This comes out clearly in the first part, where, in a characteristically elevated (and quite boring) anti-thetical, 'Roman' style, Buchan writes about pre-colonial African history and about the development of Portuguese and Dutch colonisation in South Africa. African history to Buchan is a non-history, in which there are only brief glimpses of light when an external civilisation breaks in:

> It is this overlapping of counter-civilisations, this mosaic of the prehistoric and the recent, which gives South African history its piquancy and its character. It is no tale of populous cities and splendid empires, no story of developing civilisations and conflicting philosophies; only a half-heard legend of men who came out of the darkness for a moment, of shapes warring in a mist for centuries, till the curtain lifts and we recognise the faces of to-day (Buchan 1903: 5).

Indeed, to Buchan, it is a matter for debate whether there was any progression in African culture at all, or, on the contrary, the African peoples of South Africa now represent "the nadir of a process of retrogression." What he calls the "Hottentots" and "Bushmen", at any rate,

> having within them no real vitality, have perished utterly as peoples ... [they] were fated only to prepare the way for their successors (ibid.:7).

So history and the dynamic conflict between 'civilisation' and 'barbarism' are only introduced to the sub-continent with the arrival of the European colonial powers, the Portuguese, the Dutch and the British, who each represent a progressive stage of development.

The Portuguese colonists are given a presentation which recalls Conrad's description of Kurtz in *Heart of Darkness* : Confronted with 'barbarism', their weak 'civilisation' gave in and fell back, became part of the African landscape and left only "a quaint air of decadent civilisation":

> Miserably and corruptly governed, forgotten by Europe, they forgot Europe in turn, and a strange somnolent life began of half-barbaric, wholly oriental seigneurs, ruling as petty monarchs over natives from whom they were not wholly distinct. Instead of holding the outposts of European culture, they sank themselves into the ways of the soil which their forefathers had conquered (ibid.: 27-8).

Also in more direct terms, the thin line between 'civilisation' and 'barbarism' was broken down by the Portuguese, bringing about their dissolution and collapse:

> The white man's pride died in their hearts. They were ready to mix with natives on equal terms. Now concubinage is bad, but legitimate marriage with half-castes is infinitely worse for the *morale* of a people. And since Nature to the end of time has a care of races but not of hybrids, this tolerant, foolish, unstable folk dropped out of the battle-line of life, and sank from conquerors to resident aliens, while their country passed from an empire to a vague seaboard (ibid.: 30).

The Dutch settlers, on the other hand, are a different kettle of fish, who may have their bad, brutal and primitive sides compared to British behaviour and outlook, but there is something romantic and a vigorousness about them which contrasts favourably with the flatness of the over-civilised modern European world. There is still an epic element in Afrikaner culture, whose typical historical pattern is that of the 'Great Trek':

> The thought of a national exodus comes easily to the Aryan mind,- an inheritance from primeval Asian wanderings. And in itself it is something peculiarly bold and romantic, requiring a renunciation of old ties and sentiments impossible to an over-domesticated race. It requires courage of a high order and a conficent faith in destiny (ibid.: 34).

The Boers never gave in to 'barbarism', but have kept the line dividing it from their 'civilisation' strong:

> All angles and corners, they presented an admirable front to savage nature; but they were hard dovetail into a complex modern society (ibid.: 35).

On the whole, now that war is over, the Afrikaners, if treated rightly, will provide a valuable injection of vitality into a British culture threatened with over-civilisation:

> Britain in her day has won many strange peoples to her Empire; but none, I think, more curious or more hopeful than the stubborn children of Uys and Potgieter (ibid.: 48).

The second part of the book, "Notes of Travel", is interesting mainly for its speculations about nature, but in the third part, "The Political Problem", Buchan addresses the immediate context of reconstruction, of building the ideal imperial society in South Africa, and discusses in more practical and political terms how the thin line beween 'civilisation' and 'barbarism' or 'dissolution' may be strengthened.

The most interesting chapters of this last part of the book are those that deal with the labour supply and the relations between the white and black races - the supply of cheap, African labour being the "the *sine quâ non* of mining progress" and therefore of the economic development of the future society as a whole.

The central questions are: What kinds of cultural development and progress should be stimulated among the black people of South Africa? Should "tribal" structures be kept intact or done away with? Should the long-term aim be an integration of white and black races in South Africa, and how would such an integration be imaginable? Should black people in a more short-term perspective be kept in reserves, or should they be allowed to settle in the towns to which they will attracted as labour force?

Fundamentally, Buchan thinks that the vast existing differences in development between white and black races make it irrelevant to discuss integration and a political democracy on the basis of racial egalitarianism:

> Legal rights must be granted, and ... legal equality should should follow. Social and political right must also be given; but why social and political equality? ... in this case there can be no common standard. Between the most ignorant white man and the black man there is fixed for the present an impassable gulf, not of colour but of mind ... With all his merits, the native's instability of character and intellectual childishness make him politically far more impossible than even the lowest class of Europeans (ibid.: 289-90).

Political influence for blacks must therefore be restricted to district councils, and their education must also be different - "humanism" can mean nothing to them, and therefore black education should be centered around the teaching of practical skills. The most important aspect, however, in Buchan's view is the moral problem - how will the two races be affected by their interaction?

> A coloured race living side by side with a white people furnishes one of the gravest of moral cruces. The existence of a subject race on whatever terms is apt to lead to the deterioration in moral and mental vigour of its masters ... to put it shortly, how to keep the white man from deterioration without spoiling the Kaffir, - this is the kernel of the most insistent of South African problems (ibid.: 285).

Buchan then draws a distinction between the "traditional" African, whose tribal culture is still intact, and the "unsettled" African, who provides the real problem:

... this lack of a strong, settled racial life makes it fatally easy for him to fall a victim to the vices of civilisation, and to come upon our hands as a derelict creature without faith or stamina, having lost his old taboos, and being as yet unable to understand the laws of the white man (ibid.: 286).

The solution that Buchan suggests to the problem is one of *separate development*, which on the one hand secures that Africans are not left behind in a savagery that would be a generally demoralising influence, but on the other hand also that the line dividing the races and 'civilisation' from 'barbarism' is kept intact and strengthened. A premature attempt to integrate the Africans into European 'civilisation' would bring about disastrous results:

At best the result is to turn out native pastors and schoolmasters in undue numbers, unfortunate men who have no proper professional field and no footing in the society to which their education might entitle them ... the ground must be slowly prepared before the materialist savage mind can be familiarised wih the truths of spiritual religion (ibid.: 309).

To Buchan's mind, this slow process is obviously one which will take centuries.

What we see here is, in fact, the unfolding of a theory of long-term separate development in South Africa within the British colonial administration which is completely in tune with the *apartheid* ideology for which later, Afrikaner-dominated, Nationalist governments have been made solely responsible.

In his memoirs, Buchan explained how his pleasure in mountaineering was derived from his lack of vertigo - in contrast his wife, to his great sorrow, only came with him once and became immediately dizzy: "I was fortunate to have the opposite of vertigo, for I found a physical comfort in looking down from great heights" (Buchan 1940: 134).

There is something symbolical about this statement, since Buchan also in his writings on imperialism tended to take an extremely high view, whereas his novels are full of fears of falling.[1] This goes for *A Lodge in the Wilderness* (1906) - Buchan's most general discussion of his imperialist philosophy - which is characteristically given the form of a leisurely symposium held by representatives of various typical groups within the

---

[1] As highlighted in the flying scenes in the Hannay stories.

British governing elite in a luxurious mountain lodge in Kenya overlooking the Rift Valley.

The book was written as a reaction to the parliamentary election of 1906 which ended in a Liberal victory and a set-back for imperialist positions. Buchan's intention was to gather together as many and as complex arguments as possible in favour of the idea of Empire and Imperial Federation and to argue the case for imperialism as a national unification programme that cut across traditional party boundaries, and which should be promoted by a new alliance of influential social groups.

Consequently, the book's elevated vision involves a high degree of snobbism, of Buchan indulging himself in the *mores* and rituals of upper-class socialising, as well as a generally anti-democratic attitude: Buchan is keen to distinguish the 'noble' imperialism he stands for from the vulgar, jingoistic variety of the streets, which he thinks has discredited the cause.

John Buchan returned to the discussion of the civilising mission of imperialism and of South Africa in his novel *Prester John* (1910) - another of his best-sellers which like the Hannay stories is still widely read and used as a school textbook.

The occasion for writing the book was the Bambatha rebellion in Zululand in 1906, which represented the last phase of 'primary' resistance to colonisation in South Africa, organised by members of the Zulu royal family and traditional Zulu leaders. Buchan seems to have felt that, apart from providing and exciting background for an adventure story, the rebellion was also an instance of the rule of 'civilisation' being challenged, of the thin line between 'order' and 'dissolution' threatening to break down, which called for interpretation and needed to put into the rigth perspective of understanding.

Prester John is constructed around the central thematic contradiction between 'civilisation' and 'barbarism' just as clearly as the Hannay series, but in this case with the emphasis on the social rebellion of a colonised black people as the main motive through which 'barbarism' and 'dissolution' are represented. Also the novel provided Buchan with the opportunity to bring forward in a more popular and striking form some of the central tenets from his discussion of South Africa in *The African Colony* (which had sold so poorly and been directed towards a very limited elite readership).

'Civilisation' is represented in *Prester John* first and foremost by the figure of David Crawfurd, who is an idealised Scots immigrant to the colony of South Africa. 'Dissolution', on the other hand, is incarnated by the Reverend John Laputa, a black minister who leads the rebellion against colonial rule. It is striking here that what constitutes the threat is not savage, black 'barbarity' as such, but the way in which 'barbarism' is being let loose and mobilised by a 'civilised native' - a product of missionary education.

Thus, in its overall perspective, *Prester John* seems to be an illustration of the tragic consequences to both black and white and to the whole prospect of 'civilised' development of premature racial integration - of *not* adhering to the principle of separate development that Buchan had advocated in *The African Colony*

Laputa is introduced with a certain sympathy as a grand figure, but primarily a tragic one that has to be killed off, if 'civilisation' is not to give in. The Laputa rebellion in *Prester John* provides a moral and political lesson to the white colonisers in South Africa of what constitutes their true responsibility. At the end of the story, David Crawfurd sums up the experience he has drawn from the adventure:

> ... it was an experience for which I shall ever be grateful, for it turned me from a rash boy into a serious man. I knew then the meaning of the white man's duty. He has to take all risks, recking nothing of his life or his fortunes, and well content to find his reward in the fulfilment of his task. That is the difference between white and black, gift of responsibility, the power of being in a little way a king; and so long as we know this and practise it, we will rule not in Africa alone but wherever there are dark men who live only for the day and their own bellies (Buchan 1910: 198).

Imperialism, then, in the writings of John Buchan and many of his contemporaries around the turn of the century stands out as a cultural and political programme, not so much for the spreading of 'civilisation' as for the regeneration and revitalisation of a society threatened by cultural crisis and by centrifugal forces of 'dissolution'.

The imperialist programme incorporates a range of authoritarian and anti-democratic features - both in the way it addresses the situation and problems at home, at 'the heart of the Empire', and abroad, in its ideals of colonial administration. Taken as a whole, the imperial vision is one of a monumental hierarchy of elites and masses, of rulers and workers, men and women and white and black races - a monumentalism which has many characteristics in common with totalitarian experiments in social organisation and cultural crisis management that emerge later in the 20th century.

One important difference, though, is that programmatic imperialism in its heyday around the turn of the century never becomes 'total', does not become a ruling ideology which is able to do away with all opposition.[1] At least not in Europe, where on the contrary, in spite of the efforts of John Buchan and related spirits, a development soon begins that deprives programmatic, formal imperialism of support and replaces it by more discreet varieties.

In some of the colonies, however, imperialist programmes for turning society into a monument of difference became realised and 'total' to a much greater extent. Nowhere does this apply as forcefully and blatantly as in the Union and later the Republic of South Africa, where to this day the white man preserves his "gift of responsibility, the power of being in a little way a king." And where also to this day "dark men" as well as women are kept in a position where they can "live only for the day and their own bellies."

---

[1] For anti-imperialist mobilisation cf. Kaarsholm 1987c.

Bibliography

Arendt, Hannah (1973) *The Origins of Totalitarianism*, New Edition with Added Prefaces [1st edition 1951], Harcourt Brace Jovanovich, New York.

Buchan, John (1903) *The African Colony. Studies in the Reconstruction*, Blackwood, Edinburgh and London.

Buchan, John (1916) *Greenmantle* , repr. Penguin 1983, Harmondsworth.

Buchan, John (1906) *A Lodge in the Wilderness*, Blackwood, Edinburgh and London.

Buchan, John (1910) *Prester John*, repr. Penguin 1981, Harmondsworth.

Buchan, John (1915) *The Thirty-nine Steps*, repr. Pan Books 1955, London.

Buchan, John (1919) *Mr. Standfast*, repr. Penguin 1983, Harmondsworth.

Buchan, John (1940) *Memory Hold-the-Door*, Hodder and Stoughton, London.

Buchan, Susan (1947) *John Buchan by his Wife and Friends*, Hodder and Stoughton, London.

Buckley, Jerome Hamilton (1945) *William Ernest Henley. A Study in the "Counter-Decadence" of the 'Nineties* , Princeton University Press, Princeton, New Jersey.

Etherington, Norman (1981) Imperialism in Literature: The Case of John Buchan, *Collected Seminar Papers*, No. 27, Institute of Commonwealth Studies, London.

Guillaume, André (1972) *William Ernest Henley (1849-1903) et son groupe*, Université de Lille, Lille.

Hanna, Archibald (1953) *John Buchan, 1874-1940. A Bibliography*, Shoe String Press, Hamden, Conn.

Hayward Gallery (1981) *Lutyens. The Work of the English Architect Sir Edwin Lutyens (1869-1944)*, Arts Council, London.

Herf, Jeffrey (1984) *Reactionary Modernism. Technology, culture, and politics in Weimar and the Third Reich*, Cambridge University Press, Cambridge.

Irving, Robert Grant (1981) *Indian Summer. Lutyens, Baker, and Imperial Delhi*, Yale University Press, London.

Kaarsholm, Preben (1985) Imperialism and New Journalism circa 1900, *Studies in Commonwealth Literature*, ed. by Eckehardt Breitinger and Reinhard Sander, Günther Narr, Tübingen.

Kaarsholm, Preben (1987a) *Imperialist Ideology in the Boer War Period*, Center for Research in the Humanities, Copenhagen.

Kaarsholm, Preben (1987b) Kipling, Imperialism and the Crisis of Victorian Masculinity, *Patriotism*, ed. by Raphael Samuel, Routledge & Kegan Paul, London.

Kaarsholm, Preben (1987c) Anti-Imperialism in Socialist and Working-Class Opposition to the South African War, *Internationalism in the Labour Movement, 1830-1940*, ed. by Marcel van der Linden, E. J. Brill, Leyden.

Marks, Shula and Trapido, Stanley (1979) Lord Milner and the South African State, *History Workshop Journal*, 8.

Moore, Jerrold Northrop (1984a) *Spirit of England. Edward Elgar in his World*, Heinemann, London.

Moore, Jerrold Northrop (1984b) *Edward Elgar. A Creative Life*, Oxford University Press, Oxford.

Pakenham, Thomas (1979) *The Boer War*, Weidenfeld and Nicolson, London.

Parry, Benita (1983) *Conrad and Imperialism. Ideological Boundaries and Visionary Frontiers* , Macmillan, London.

Pyrah, G. B. (1955) *Imperial Policy and South Africa, 1902-10*, Clarendon Press, Oxford.

Schröder, Hans-Christoph (1978) *Imperialismus und antidemokratisches Denken. Alfred Milners Kritik am politischen System Englands*, Institut für europäische Geschichte, Mainz, Vorträge Nr. 68, Franz Steiner Verlag, Wiesbaden.

Searle, G. R. (1971) *The Quest for National Efficiency. A Study in British Politics and Political Thought, 1899-1914*, University of California Press, Berkeley and Los Angeles.

Semmel, Bernard (1960) *Imperialism and Social Reform. English Social-Imperial Thought 1895-1914*, Allen & Unwin, London.

Smith, Janet Adam (1965) *John Buchan. A Biography*, Rupert Hart-Davis, London.

Smith, Janet Adam (1979) *John Buchan and his World*, Thames and Hudson, London.

Thompson, L. M. (1960) *The Unification of South Africa, 1902-1910*, Clarendon Press, Oxford.

Orientalism and the Ancient Near East

Mogens Trolle Larsen

Herder's Dream of the Past World

Between 1784 and 1791 Johann Gottfried Herder published his great work *Ideen zur Philosophie der Geschichte der Menschheit*, in which he presented a vast synthesis of all that could at his time be known about the history of the entire Globe. It was a very interesting time for such a summary to appear, for only a few decades later the European exploration of many areas, such as the Near East would begin to provide real knowledge that could replace or at least embellish the fund of received traditions on which Herder largely had to build; for the worlds of the ancient Near East these were constituted primarily by Greek and Roman literature and by the Old Testament.

Herder dedicated his chapter 12 to the Near East, and at the end of it, just before turning to the much more clearly illuminated Greek world, he wrote:

> With the sorrow of a wanderer who must leave a land before he has come to know it as he wished, I have to leave Asia. How little it is that we know of it! And mostly from how late a time, from such uncertain hands! .. In the Near East and neighbouring Egypt everything from the ancient times appears to us as ruins or as a dream which has disappeared. .. The archives of Babylon, Phoenicia and Carthage are no more; Egypt had withered practically before the Greeks saw its interior; thus, everything shrinks to a few faded leaves which contain stories about stories, fragments of history, a dream of the world before us (Herder 1784-91: 329).[1]

---

[1] Mit dem Bedauern eines Wanderers, der ein Land verlassen muss, ohne daß ers nach seinen Wunschen kennenlernte, verlasse ich Asien. Wie wenig ists, was wir von ihm wissen! Und meistens aus wie späten Zeiten, aus wie unsicheren Händen! ... Im Vorderasien und dem ihm nachbarlichen Ägypten erscheint uns

Herder did have a premonition that the archives of Babylon might not be irretrievably lost, however, for travellers such as Carsten Niebuhr had brought back hints that the massive ruins contained "Tafeln chaldäischer Schrift". On the other hand, Herder had little confidence in the ruins themselves, for he suspected that the buildings and the art to be found there would turn out to be uninteresting.

A century later Herder's dream of opening up the old mounds was being realised, and monuments and texts were being carried in enormous quantities to Europe where they were on display in the major museums. 1884, the centenary of Herder's book, was also the year of the Berlin Conference where England, France and Germany divided up the African continent among themselves, culminating in the "Scramble for Africa" which in some ways resembled the competition among European nations for access to the most promising ruins and consequently the most impressive monuments and antiquities for their museums. It seems reasonable to inquire about the motivations behind the great effort which had been and which was at that time with increasing intensity being expended upon the exploration of the past of the Near East.

Herder himself gave one powerful motivation for the sacrifice and effort which was put into the archaeological exploration of the ancient civilizations: to throw light upon the origins of the cultural complex which developed over the millennia into our own European civilization.

Even though he was writing at a time when practically nothing certain could be said of the ancient worlds, it was nevertheless clear to Herder that it was in the Near East that we should find the origin of innumerable human achievements in culture and technology. The "Traum der Vorwelt" pointed directly to a past which was also in a sense the origin or precursor of our own:

> We wander around so to speak on the graves of fallen monarchies and see the shadows of their former activity on the earth. And this activity has truly been so great, that if we count Egypt to this area, no other region of the world apart from Greece and Rome has invented and prepared so much, especially for

---

aus der altern Zeit alles wie eine Trümmer oder wie ein verschwundener Traum ..... Die Archive Babylons, Phöniziens und Karthagos sind nicht mehr; Ägypten war abgeblühet, fast ehe Griechen sein Inneres betrachten; also schrumpft alles in wenige verwelkte Blätter zusammen, die Sagen aus Sagen enthalten, Bruchstücke der Geschichte, ein Traum der Vorwelt .

Europe, and through it for all nations of the world (Herder 1784-91: 302).

In fact, this was a strong motive for the allocation of generous funds for excavations as well as for the creation of large collections of antiquities in museums in the West, and later the setting up of academic posts, research programmes and education systems in the universities.

One major aspect was the idea of the Cradle of Civilization, the starting point for the historical development or evolution which ended with Europe and the West. In the period of Western hegemony over large parts of the Globe, of colonialism and imperialism, a unilinear view of world history which marked out the western civilization as the concluding glory of millennia of development, clearly had a lot of attraction. Another exciting aspect was the possibility of finding information which could throw light upon, explain or perhaps show parallels to European religion as documented in the Old or maybe even the New Testament.

A history of Assyriology and Near Eastern archaeology in the nineteenth century cannot, of course, be presented here, but suffice it to say that around 1884 these were very active disciplines which were developing rapidly in the various major European countries (Budge 1925; Lloyd 1955 and 1963). Excavations had started in the 1840s when French and British expeditions began to tackle the immense ruins of the major Assyrian capital cities in the north of Mesopotamia, finding palaces filled with wonderful reliefs and majestic bull colossi which had guarded the various important doorways. In these ruins were also discovered very large collections of clay tablets which were brought back to Europe together with the major part of the sculpture. By 1860 it seems fair to say that the initial phase of decipherment was over and the scholars at the Louvre and the British Museum could begin to read all these texts and try to make sense out of their contents.

Under the rubrics 'Nationalism' and 'The Bible' I shall discuss two aspects of the growth of Assyriology: its relationship to the political realities of the time, colonialism, nationalism and anti-Semitism; and its relationship to other academic disciplines, primarily theology and the history of religion. Obviously, all of these considerations are tightly linked and my analysis therefore introduces distinctions which are to some extent artificial, but I hope to be able to show how even such an apparently unworldly and innocent academic discipline was connected in different ways with the dominating ideological concerns of the period.

I have chosen to concentrate my attention on the developments in Germany - in order to make the task manageable, and because Assyriology had its most dramatic battles here during the period in question.

Nationalism

Nationalism was as we know a political reality of the greatest importance in Europe of the nineteenth century, and our academic disciplines were of course deeply concerned with matters of national prestige and ideology. In the very earliest phase of archaeological exploration in Mesopotamia the French and English diggers frequently fell out with each other over rights to ruins or mounds which appeared particularly promising, and the Sublime Porte in Istanbul was being appealed to by ambassadors and special envoys - for it was of course the Ottoman empire which was supposed to exercise authority in the region.

Bismarck's strong new German state, which was looking for "a place in the sun" and establishing itself as a world power with a colonial empire, had not been involved to any appreciable extent in the excavations in the Near East during the first phase in the nineteenth century, but the desire to participate directly was very strong. As with the Scramble for Africa (see Betts 1966), it seems that there was a strong element of the "So ein Ding müssen wir auch haben" feeling in the German interest in the ancient Near East, but it should be remembered that Germany was developing a very substantial interest in Turkey.

Plans for ventures in the region were accordingly supported at the very highest level of German society, as is apparent from the formation of the Deutsche Orient–Gesellschaft in January 1898. In the Egyptian pillar hall of the Museum in Berlin a distinguished group of 60 gentlemen assembled to set up the new association. Prince Heinrich von Schönaich-Carolath was in the chair and introduced the meeting with a speech in which he remembered Goethe's dictum "Gottes ist der Orient, Gottes ist der Occident"; he expressed the hope that Germany would never be missing in personalities who were willing to follow ideal goals, and the association was set up as an official body with two presiding officers, princes Carolath and Alexander zu Hohenlohe–Schillings-fürst; the executive officer was Admiral Hollmann.[1]

The "Aufruf" which had been sent out to authorities all over Germany to explain this new society pointed out that the burning interest in Biblical matters had created a great public interest in the finds from the Near East especially in the Anglo-Saxon world, and added that:

> The undersigned are of the opinion that the time has come for Germany to take part in the great task of discovering and recovering the earliest Orient through more extensive systematic excavations (OLZ 1898: 36).

In a debate in the Prussian Landtag a couple of months after this meeting it was pointed out by Herr minister Bosse that at the very moment a German scientific expedition was exploring sites in Assyria, and he agreed that these activities "hold out the promise of quite substantial advantages for the future" (OLZ 1898: 124).

It seems relevant to wonder what such advantages could be. I have no doubt that the minister had several different aspects of the situation in mind. The prestige of participating on an equal footing with England and France in the exploration of the world, and the glory of great antiquities in the national museums certainly played a considerable role. The acquisition of antiquities did not at that time cause any moral scruples, of course, and it was remarked during the debate in the Prussian Landtag that it was only through the extreme courtesy of English and French scholars, who gave a few antiquities and casts of great finds to the Germans, that it was at all possible to study ancient Egypt in Germany. On the other hand, "of course they still take the lion's share for themselves". It is interesting to note that the same member of the Landtag put in a special plea for the resumption of the excavations of the site of Sendschirli in Syria, stressing the point that they were "nur nationale Arbeit: daran hat niemand mitgearbeitet, der nicht zu uns gehörte" (OLZ 1898: 123).

---

[1] See the report in Orientalistische Literatur-Zeitung (OLZ) 1, No. 2, 15. February 1898, 33-38, which is based on an article in "Norddeutsche Allgemeine Zeitung" of 25 January 1898. A scientific "Beirat" was also set up, including representatives from the Kultusministerium, the Museum and the Academy of Sciences plus five professors: Delitzsch, Zimmern, Sachau, Conze and Lehmann.

But perhaps we should also listen to the British Assyriologist Sir Wallis Budge who wrote a history of the discipline, and who had this to say about the motivations behind the German efforts:

> and as for excavations in Assyria and Babylonia, many shrewd observers have remarked that Germany only began to excavate seriously in those countries when she began to dream of creating the German Oriental Empire, which was to be reached by way of the Baghdâd Railway (Budge 1925: 293).

For a British gentlemen writing in 1925, fully aware of English domination of the area, this German dream could only be an object for ridicule, but in 1898 it could still be dreamt.[1] And Budge's book itself is bound up completely in the nationalistic fogs of the time: his aim was to prove beyond doubt that the discipline of Assyriology was created by Englishmen. Therefore, whereas he can gracefully concede that "Assyriology in its early stages owed much to Denmark", he hastens to add: "and very little to Germany"; and he muses:

> it is a curious fact, but the German Assyriologists with whom I came into contact in the British Museum showed, by their talk and behaviour, that they believed that the science of Assyriology was founded by the Germans, and that they had taught the rest of the world how to decipher and translate the cuneiform inscriptions (Ibid.: 293-295).

By 1880 German scholarship had a very high standard indeed, and it is symptomatic that when American universities in the period wished to set up programs in research and teaching in Assyriology, they imported scholars from Germany (Meade 1974). But the strong emphasis on philological excellence, exemplified by the two scholars who stand as the founders of the discipline in Germany, Eberhard Schrader and his pupil Friedrich Delitzsch, perhaps gave German Assyriology a somewhat peculiar flavour. Budge, in his rather crude but sometimes very funny accounts of German scholars who came to the British Museum to study the texts excavated by British archaeologists, points out that it was always British or French scholars who had to copy the texts and

---

[1] See the title of a pamphlet published in 1886 by a certain A. Sprenger: "Babylonia, the richest land in ancient times and the most promising colony for today" (Sprenger 1886).

prepare the first editions; only after this basic work had been completed the German scholars would come along with their criticism, would prepare new, revised text editions, and pretend that the original English or French scholarship had never existed. And in the rare cases where German scholars did produce an editio princeps of a text, it was always of a low quality. He adds:

> This statement is not intended to detract in any way from the value of German work on the minutiæ of grammar or on points of philology, which are often of the greatest importance, or the German faculty for elaboration and patient research, which is beyond all praise; but German Assyriologists were never good copyists (Ibid.: 292).

There were other explanations, of course, one of them suggested by the great Berlin Assyriologist Hugo Winckler.[1] His reasoning stands as a curious echo of the colonialist dream of a place in the sun, since he felt that German scholarship suffered from the fact that Germans generally had been excluded from active contact with the world, so that the scholar had only been able to get to know the world through his work in the "Studierstube". "Die berühmte Unpraktischheit des deutschen Gelehrten" could be directly explained by this, and this isolation had furthermore had the effect of making philology and linguistics the main concern of German scholars, whereas the understanding of other cultures and peoples was a poorly developed field where Germans had very limited competence (Winckler 1906: 10).[2] But things were changing, and for a short while before the First World War Germany gained for itself the coveted "place in the Sun". In 1906 Winckler went to Anatolia to start excavations of the Hittite capital at Boghazköy, and he was in fact typical of the situation in Germany, for the scholars there had initiated the great tasks which placed Germany in the forefront of Near Eastern archaeology. 1899, the year after the founding of the Deutsche Ori-

---

[1] Characterised by Budge as follows: "he was somewhat of an erratic genius; and though his learning was considerable, his deductions were frequently unsound, and many of his theories were based on his imagination" (Ibid.: 234).

[2] As an aside let me tell you that Winckler had little patience with German scholars who followed the British lead and dealt primarily with the tablets; his review in 1898 of a catalogue of texts in the British Museum by his German colleague Bezold contained the devastating note that there are types of work: "die ihre Ziele durch Schonung des Hirns auf Kosten anderer Körperteile erreicht" (OLZ 1898: 57).

ent-Gesellschaft, excavations started in the ruins of Babylon, and in 1903 extensive work started at the site of the ancient northern capital Assur.

The Bible

In the memorandum written as the background paper for the Deutsche Orient-Gesellschaft it was stated that research in the remains of the ancient worlds of the Near East had given a new understanding of

> those original conditions .. where we find the roots of our own culture, our time-reckoning and starlore, our system of weight and measure, as well as important parts of the religious concepts which are contained in the Old Testament (OLZ 1898: 36).

As the texts and monuments were beginning to be understood and interpreted, it became clear that here was a great potential for penetrating deeper into a comparative study of biblical traditions, and in fact, the discoveries in many respects demanded a revision of cherished and extremely firmly established doctrines.

Public interest in such questions was clearly very strong - as became evident in England in 1872 when George Smith, while working on a publication project in the British Museum, discovered a fragmentary text which he found to contain an Assyrian version of the Story of the Flood, what became known as the "Deluge Tablet". The discovery was made public when Smith read a paper before the Society of Biblical Archaeology on December 3, 1872, and it created a sensation in England and France; present on the podium was a group of distinguished men and women, headed by the Prime Minister W.E. Gladstone personally. Less than two months later Smith was on his way to Nineveh where he was supposed to find the rest of the tablet, and his expedition was financed by the newspaper Daily Telegraph which had given a grant of 1000 guineas to the British Museum for this express purpose. A week after his arrival at Nineveh he had the good fortune to discover another piece of the text in the dump left by the old excavation (Budge 1925: 112-114; Lloyd 1955: 180).

We are here dealing with questions which at the time, in the religious world of fin-de-siècle Europe, were far from trivial and which were of direct interest to very large audiences. Budge remembered that "Mr. Basil Cooper, a Press correspondent, used to visit the Museum every week to obtain for his paper information of any Biblical parallel that had been discovered since his last visit" (Ibid.: 271). However, in the world of biblical scholarship the new discoveries had a quite limited impact for a long time.

The school of Wellhausen dominated the German theology of the time, and its emphasis on a strictly literary/critical approach to an analysis of the Old Testament left little space for the new, predominantly historical evidence. The prevailing view of both the theological professors and the general public was that the monotheistic religion of the Jews was to be seen and understood as a truly unique phenomenon, growing directly out of the nomadic, primitive world of the tribes. Early Arabic poetry was brought into the analysis to further an understanding of a social system which was supposed to have formed the basis for the new religion (see Kraus 1969, especially chapters 9-11).

In that situation the Assyriologists argued with increasing intensity that Israel was only a small part of a very large world, that there was constant interaction and contact throughout the entire region, and that consequently (in the words of Hugo Winckler):

> neither Arabia nor Kanaan could entrench themselves behind Chinese Walls and keep out the air which the entire Orient has breathed through centuries (Winckler 1906a: 15).

So the very first thing that Assyriology could do was to provide a background for the Old Testament, placing it in a historical, cultural and religious context whose very existence had been unknown.

German Assyriologists were eagerly pursuing questions of this nature all through the period in question, and several schools developed which sought in different ways to place the Near Eastern evidence in relation to both the Old and the New Testament. The father of Assyriology in Germany, Schrader, as early as 1863 published a study of Biblical "Urgeschichte", and after that a whole series of books appeared which tried to present the ancient texts in a way that was meaningful in terms of Bible studies. Heinrich Zimmern, one of the most influential Assyriologists of the time, published a volume which in its title declared Assyriology to be an auxiliary discipline for the study of the Old Testament and the classical world (Zimmern 1889).

Of the explanatory schemes which were suggested at the time I have chosen to concentrate on two: Pan-Babylonism and the debate known as "Babel und Bibel".

The Pan-Babylonist scheme is linked to the name Hugo Winckler. He was a very active participant in the debates concerning Assyriology and the Bible, claiming for himself the position as a historian and definitely not a theologian. He wrote a History of Israel and had a considerable influence on a number of theological scholars, even followers of Wellhausen. His grand scheme which is known as "Pan-Babylonismus" was based on the idea that all myths everywhere can be seen as reflections, mostly distorted, of a system which was developed in Babylonia around 3000 B.C. This system was an astral religion in which the sun's path through the Zodiac, the rising of the Pleiades, and the Precession played a central role. The Mesopotamian origin of all such features around the Globe could be seen from the fact that this was the only place where the whole theory was ever understood (Winckler 1901, 1905, 1907).

The immensity of the claims put forward by Winckler appears for instance from the title of one of his books which would be as follows in English: "The Babylonian view of the heavens and the universe as the basis for the world conception and mythology of all peoples" (Winckler 1901). It is therefore understandable that when he started a new scholarly journal, he chose the name "Ex Oriente Lux" for its title. Naturally, this scheme came under attack from many sides, but Winckler's immense knowledge and very sharp pen kept the issue alive at least until around 1910.

The Bible was of course to be understood by way of this scheme as well, and Winckler was adamant in his claim that the fundamental ideas in Biblical religion were dependent on the nature of the ancient Near Eastern religions. Perhaps his most astonishing claim here was that these Mesopotamian systems of belief were based on the concept of one mighty divinity who stands over or behind all of the numerous divine manifestations (Erscheinungsformen):

> the basic idea in this polytheism is therefore not at all that for instance many gods have created the world, i.e. have embodied themselves in it, but that one great divine power shows itself as acting through innumerable forms of appearance. But of course - this is the teaching for those who have been initiated, those who know; only the unending diversity of the forms of appearance was communicated to the people through the cult and the mythology (Winckler 1906b: 78)

This knowledge was shared only by the knowing few, but it is implied that Abraham (who was after all a Babylonian, said Winckler) was among that group of sages - and he was the father of the Jewish religion. However, his faith came to represent a reaction against the Mesopotamian practices by making the demand that

> not only the priest or the initiated must realise the actions of a transcendent deity behind the statue of a god or the heavenly bodies in the Universe, but everybody must be aware of this (Winckler 1907: 133).

Therefore the ban on representations of the manifestations of the deity and the stress on the singularity of God. Nevertheless, the Jewish people had their own beliefs and had to be constantly kept on the right path; from this Winckler concluded:

> People and religion are not identical, the religion did not grow out of the Israelite/Judaic nationality, but was brought to it through a higher spiritual culture, in the same manner that other religions were brought to other peoples. The Biblical religion is not a creation of the people of Juda/Israel, but Jewry is a creation of the Biblical religion (Winckler 1906b: 19).

This is perhaps an instance of what Hans-Joachim Kraus calls "teilweise maßlose Selbstüberschätzung" on the part of the Assyriologists (Kraus 1969: 306), and it was not unique in this respect.

Winckler had asked the uncomfortable question whether the entire history embodied in the literature from the ancient worlds would not dissolve into mythology, when more and more stories relating to "historical" figures could be seen to reflect basic mythological structures (Winckler 1906a: 17). This danger became even more apparent in the work of another influential scholar, Peter Jensen, who developed a majestic scheme which had some resemblance to Winckler's. Jensen wanted to destroy the New Testament as a religious text, as well as the Old, and his basis for some very extraordinary claims was his interpretation of the Babylonian Epic of Gilgamesh. This text, composed in its canonical form sometime round the middle of the second millennium B.C., but which goes back to a series of Sumerian myths and stories about a great king who ruled in the city of Uruk, is the classic literary text from ancient Mesopotamia. For Jensen it was the source of all mythical motifs in world literature - Indian, Greek, and of course also Hebrew. In line with the prevailing mode of explanation at this time the epic was to be understood as based on astral conceptions, of course, with the hero Gilgamesh playing the role of the sun. Jensen suggested that the stories about the major figures in the Old as well as the New Testaments had to be understood as reflections of the story told about the old hero, and he argued with great passion that we should free ourselves of this old myth:

> Yeah! Jesus of Nazareth, in whom the Christian world has believed for at least two millennia, and perhaps even much longer, as the Son of God and the Saviour of the world, and in whom even the most advanced science of our day sees at least a great human being who once wandered on Earth and died as a high ideal and model - this Jesus has never wandered on Earth, has never died on Earth, because he is simply nothing but an Israelite Gilgamesh. ... We, the children of a time praised for its wonderful cultural accomplishments, and who like to look with pitying smiles at the beliefs and customs of peoples of the past, we serve in our cathedrals and houses of prayer, our churches and schools, in palace and hut, a Babylonian god, Babylonian gods! Yeah! we, a great, powerful cultured people of German brethren, we tear each other asunder in a fight over the tinsel which has been hung around the neck of this foreign god; we block the sources of the power of our people, because a man sits on the throne in Rome, who is the vicar of this Babylonian sungod - purely on the basis of human custom and human whims (Jensen 1906: 1029-1030).

Jensen's enormous book in which this message was explained with exquisite attention to philological detail, appeared in the middle of a storm which had been sweeping Germany since January 1902, when Jensen's teacher, Friedrich Delitzsch had held a lecture in Berlin with the title "Babel und Bibel".

Delitzsch, the son of the famous Evangelical theologian Franz Delitzsch, who was at that time at the peak of his career, was invited to give a series of lectures to the Deutsche Orient-Gesellschaft after his return from an exploratory journey to the Near East. The first of these was delivered in the hall of the Singakademie in Berlin on the 13th of January 1902, and the prestigious nature of the occasion was evident from the fact that His Majesty the Emperor Wilhelm was present. The importance of the topic was further emphasized by what followed three weeks later, when Delitzsch was invited to repeat his performance at the imperial palace in Berlin where the Emperor heard it once more in the company of his queen and members of the court. And it became clear immediately that it was not just Wilhelm who had an antiquarian interest in Assyriology, for Delitzsch's lecture started a flood of newspaper polemics, pamphlets and books which lasted for several years.

Delitzsch began his lecture by squarely addressing the question of the motivation for the whole enterprise: Why all these efforts in distant, inhospitable and dangerous lands, and from where comes the public interest in the excavations? The answer, he said, was The Bible.

In his lecture he then proceeded to give examples of the wonderful new discoveries, and he showed how in several ways these could throw completely new, and in some instances quite unexpected light on the Old Testament. However, he insisted that Babylonia was a truly great civilization which had dominated culturally in the entire region of the Near East, and so one should not just look for or expect to find parallels to Old Testament stories and practices, for in fact, Babylonia was bound to be the origin of them (Delitzsch 1902).

The circumstances surrounding Delitzsch's lecture guaranteed a wide public interest of course, but they can hardly explain the fervour with which it was greeted, in Germany as well as in practically all other European countries. It was translated into several languages and pamphlets attacking or supporting his positions appeared within a few months both in Germany and abroad.[1]

And yet, Delitzsch's colleagues universally claimed that he had really said nothing new, that it had all been known and discussed in assyriological and Old Testament scholarly circles. The important thing was that now the general public was drawn into the debate, and the specific claims made by Delitzsch made it imperative for theologians to respond, either publicly for or against Delitzsch, or privately by incorporating the ideas into their own understanding of the religion they adhered to.

---

[1] The German debate was covered extensively in regular reports in e.g. the journal *Die Christliche Welt. Evangelisches Gemeindeblatt für Gebildete aller Stände* 1902: 188-189, 943-946, 1042-1044; 1903: 227-229, 243-246, 275-278, 491-493, 539-542, 797-799, 1034-1036; 1904: 62-64, 294-297, 638-640; 1905: 155-157, 446-448, 1112-1114, 1238-1240.

One of the very first reactions of any substance came already the next month in the weekly "Die Christliche Welt" where a repeat performance of the lecture in Frankfurt was reviewed. The article deserves to be quoted at some length because it gives the sense of excitement and something close to awe which obviously inspired many of the people who listened to Delitzsch:

> a deep impression was created by the compelling force of the reasoning by which he - there is no other word: *proved* the dependence of Israelite culture and religion on Babylonian-Assyrian phenomena - by way of a presentation and clarification of a rich wealth of irrefutable facts. And truly heartwarming was the passion with which the lecturer presented as the profit of the excavations and researches the liberation of the true religion, as we know it from the Prophets and from Jesus, from human additions which - whatever their poetic value - constitute an unbearable burden for the conscience of modern man (Foerster 1902: 189).

Precisely this, it seems to me, was the message in Delitzsch's first lecture: the Christian faith is freed from the weight of the traditions which in the word of the reviewer had become "an unbearable burden for modern man."

In the second lecture, held a year later in the same prestigious surroundings with the Emperor and his queen present, Delitzsch made this point very clearly indeed when he claimed that the results of his research:

> meant, it seems to me, the end of the theological approach to the Old Testament and the beginning of new one, based on the history of religions (Delitzsch 1903: 44).

Delitzsch took the radical stand to deny the entire Old Testament its nature as a divinely inspired book, a revelation. This meant also the books of the Prophets, of course, and that was too much for most Christians - and certainly for Kaiser Wilhelm. He took the really quite momentous step of writing an open letter addressed to Admiral Hollmann, the president of the Deutsche Orient–Gesellschaft,[1] in which he denounced Delitzsch's stand and explained what he felt was the correct position to take. Delitzsch

---

[1] Published in the journal *Die Grenzboten. Zeitschrift für Politik, Literatur und Kunst,* 62, Nr. 8 from 19. February 1903.

ought to concentrate on Assyriology, where he was an expert, and stay away from theology where he was an amateur.

It should be kept in mind that there were at least two debates involved at the same time. Delitzsch brought forward a number of interpretations of Mesopotamian texts and traditions, and he certainly did not convince all of his Assyriological colleagues of such matters as the correct interpretation of the term *shapattu*, which he compared to the Sabbath-institution, or of his claim that the Hammurapi-dynasty was "Canaanite". The debate about such problems could not, however, be kept apart from the wider discussion of the question of the proper approach to the Old Testament.

However, there was a further strand in Delitzsch's writings which to a modern eye seems particularly important, but which was not dealt with, at least not directly, in the scholarly debate to which I have so far had access, the critique of Jewish religious ideas which had a clear relationship to the anti-Semitic currents in Germany. In his Notes to the second lecture Delitzsch expressed surprise at the hostile reception his ideas had been granted by Jews, and he claimed that the discussion started by him could not possibly be regarded as an attack on or insult to the modern Jewish faith (1903: 41-42).

But it could not be denied that Delitzsch's denunciation of the Old Testament as "eine Offenbarung" also meant an all-out attack on the Jewish faith, of course. A central point related to his claim that the Jewish monotheism was not a truly universal faith, since Jahweh was a "national deity", tied to "his people", somewhat in the same way that the god Assur was a national god. Delitzsch regarded the Jewish religion as an example of a "national–partikularistischer Monotheismus", and he held out Jesus and the Christian faith as the first universal religion. Involved in this was his strongly worded emphasis on "Weiterbildung der Religion" which ended his second lecture, and which may have been the direct reason for the intervention from the Kaiser.[1]

In the third lecture, held in October 1904 before the literary societies in Barmen and Cologne, in rather harsh contrast to the glory of the first two audiences, Delitzsch spent a long time praising the virtues of Babylonian moral and ethical norms, comparing with the Old Testament and not finding a great difference. He

---

[1] See *Die Christliche Welt* 16, Nr. 9, 1903, 212.

ended with a passionate praise of Jesus and the Christian faith, and his last words were:

> thus we, who study the old Babylonian world and watch how the leading minds in Babylonia were concerned in serious eagerness, indeed, in fear and trembling, to seek God and The Truth - we can only happily accept that the Evangelist lets the Wise Men from Babylon be the first to bring their homage to the Cradle of Christianity (Delitzsch 1905: 48).

There can be no doubt that Delitzsch's lectures inscribe themselves in the strongly anti-Semitic feelings which existed in Germany - as in most other European nations at the time. His position became even clearer when in 1921 he published the book "Die grosse Täuschung", or "The Great Deception", which was his name for the Old Testament. This book, he declared, was entirely superfluous for the Christian church and the Christian family, and he suggested that instead of turning to it:

> we should from time to time become absorbed in the deep thoughts which our German heroes of the mind have thought on God and the Hereafter and Immortality, such as they have been so excellently collected in Wilhelm Schwaner's Germanen-Bibel (Delitzsch 1920-21, part 1: 95).

This, of course, was what happened under the Third Reich with the "Deutschen Christen", and the final argument is logically that Jesus was not only a non-Jew, he was in fact an anti-Jew:

> and who still wants to stick to the constantly stated, though entirely unproven mistake that Jesus belonged to the Jewish race, let him acquaint himself just once with the abysmal contrast between the passionate speeches of any of the Prophets of the Old Testament and the wonderful, truly heavenly calm that breathes through the parables of the Sage of Nazareth! A greater inner and outer contrast is not conceivable (Delitzsch 1920-21, part 2: 68-69).

With inexorable logic Delitzsch reminds us that the sentimental cliché: "the Jews have given the world its saviour" should be replaced by the less dubious statement: "The Jews have killed the Saviour of the World" (Ibid.: part 1, 94).

Conclusion

The men whose work I have discussed here were great scholars, who did wonderful scientific work in the discipline, work which is still valid. They were involved in a field of study which in different ways had to challenge cherished beliefs and positions, and their discoveries were of potential interest to emperors, princes, prime ministers, university professors - and the lay public. They were deeply concerned about the right position of their new discipline in the intellectual establishment, and they clearly felt that they had something absolutely revolutionary to contribute. Intense scorn was heaped on the conservative theologians and priests who would not or could not understand what had happened. It was of course a problem that the Assyriologists could not quite agree among themselves what it was, but at least they were searching (see Finkelstein 1958 and Ebach 1986).

The grandeur of the claims put forward meant that the Assyriologists were bound to get into open conflict with other scholars, not just from Theology, but from many different disciplines. The great Viennese ethnologist Wilhelm Schmidt held a lecture in 1908 where he defended the concept of "Elementargedanken" against the Pan-Babylonian scheme. Basically, Bastian's idea of "Elementargedanken" was an anti–diffusionist theory which held that basic cultural traits would and could develop independently in different cultures, and it is clear that if the claims brought forward especially by Winckler were to be upheld, it would have very serious repercussions, so we see the disciplines beginning to defend themselves. Schmidt says with some horror:

> The entire discipline of Ethnology, or at least the part of it which deals with mythology and religion would .. simply be made into a dependency of Assyriology, and no one would in future be a complete Ethnologist without having completed his course in Assyriology (Schmidt 1908: 75).

Clearly, that would not do! Classical scholars on their part could not be very happy about the situation either, and there is a curious little pamphlet from 1904, written by a professor of Philosophy at Berlin university, which addresses this. I cannot refrain from quoting a small section:

> There is a remarkable difference between us and the Assyriologists: we believe that we are able to impress satisfactorily

even the 20th century through our science and above all the facts themselves; but they - granted that everything in Babylon is inferior compared with Greece and Rome, and that the Assyrians were an unspeakably abominable people - they fear not to be able to create sufficient interest without sophistries (Blass 1904: 19). [1]

So, the Assyriological sophistries should obviously simply be forgotten. And that is what happened very quickly: Winckler's Pan–Babylonian fantasy was cruelly demolished in 1910 by Kugler, who as a Jesuit, an astronomer and an Assyriologist was in a position to demonstrate the complete lack of convincing evidence in favour of Winckler's astral religion (Kugler 1910). Jensen's dream of Gilgamesh just died. Friedrich Delitzsch died in 1922, and his ideas became part of the general craziness.

The debate is of interest today as a significant element in the ideological conflicts of fin-de-siècle Germany and as an example of how a seemingly innocent academic pursuit was involved with and shaped by the intense political concerns of the age - the period when in fact most of the academic fields were either created or at least received the methodologies and basic characteristics which are still very much alive.

The assyriological fights of this period certainly have a direct interest for the scholars in Assyriology and ancient Near Eastern studies, since they have had a profound and lasting impact on the concerns and practices in these disciplines. This was pointed out already in 1925 by Budge, and his fears have shown themselves to be quite realistic:

> Indeed, it is possible that some of these systems have wrought permanent injury. Thus scholars who do not believe in "Pan-Babylonismus" have been led, by a natural bias, to underrate astrological influence in the Babylonian religion. And many who recognized it have been chary of drawing natural and legitimate inferences because of their possible misuse in the hands of others (Budge 1925: 283).

Such questions as the astral elements in Mesopotamian mythology and religion have simply been dropped by the discipline - nobody wants to become associated with ideas of that nature, even though it is often not understood at all why these topics

---

[1] See OLZ 1904: 121-125.

should be avoided. And the relationship between Mesopotamian and Old Testament cultural and religious concepts can hardly be said to constitute a flourishing area of study - even though it is obvious that the final word has not been said about very many problems in this area.

It is not surprising that religion became a major element in the practice of the discipline, since at the time the logical framework for an understanding of the monuments and texts from Mesopotamia had to be the Old Testament. From such concerns grew a complex and sometimes confusing debate which attempted to provide a basis for an understanding of the relationship between Jewry and Christianity, a debate which of course was located in the political discourse of the time. Anti-Semitism was certainly not invented by Assyriologists, it was an ideology which offered a conceptual structure that some scholars found appealing and which came to have disastrous consequences for the development in German Assyriology. I have no reason to believe that in the thirties the discipline had a larger number of Nazis in its ranks than other academic fields, but the question of anti-Semitism became a burning concern there at a very early time because the scholars in their daily practice were confronted with issues that could so easily be placed in that context.

The ideological intensity which characterised fin-de-siècle Europe is clearly evident in the Assyriological battles and some of the protagonists were described as "von extremer Radikalität". Hugo Winckler was known to be violently anti-Semitic, even though his writings in general seem to be more subtle than those of Delitzsch; and the latter can hardly be described in any other way, even though he defended himself vigorously against the charge of anti-Semitism - "that slogan which serves as a welcome cover for the anti-German and anti-Christian attitudes of a lot of accusers". He claimed that his work was located in the space between philo- and anti-Semitism, and that "it will have to be recognized that I am guided only by *incorruptible love of the truth* ('unbestechliche Wahrheitsliebe')" (1920-21, part 2: 4).

As a Christian and being involved with questions which belong in the border area between Old Testament studies and Assyriology, Delitzsch clearly felt that no compromise was possible, and in fact, you cannot - according to Hans-Joachim Kraus - remain neutral:

The critic is always, and already through the process of his research placed in a crisis in which the question of Truth
demands an answer from him (Kraus 1969: 313-314).

In this passage Kraus echoes the pathetic and constant use of the
word "Wahrheit", so characteristic of the writings discussed here,
but the concern with this term was not a peculiarly Assyriological
phenomenon. The Truth, not one of many possible truths, not a
factual synthesis, was a goal that could and should be resolutely
pursued, a dangerous dream which was in fact characteristic of
the passionate desire of the age to create a new century which
should be based on revolutionary, even if also often painful and
unpopular truth.

Bibliography

Betts, Raymond, F. (1966) *The "Scramble" for Africa. Causes and Dimensions of Empire.* Problems in European Civilization. D.C. Heath and Company, Boston.

Blass, Friedrich (1904) *Wissenschaft und Sophistik.* Vaterländische Verlags- und Kunstanstalt, Berlin.

Budge, Sir E.A. Wallis (1925) *The Rise & Progress of Assyriology.* London.

Delitzsch, Friedrich (1902) *Babel und Bibel. Ein Vortrag.* J.C. Hinrichs, Leipzig.

Delitzsch, Friedrich (1903) *Zweiter Vortrag über Babel und Bibel.* Deutsche Verlags-Anstalt, Stuttgart.

Delitzsch, Friedrich (1905) *Babel und Bibel. Dritter (Schluss-)Vortrag,* Deutsche Verlags-Anstalt, Stuttgart.

Delitzsch, Friedrich (1920-21) *Die Grosse Täuschung* 1-2, Deutsche Verlags-Anstalt, Stuttgart und Berlin (part 1) and Karl Rohm in Lorch, Württemberg (part 2).

Ebach, Jürgen (1986) Babel und Bibel oder: Das "Heidnische" im Alten Testament, in Richard Faber and Renate Schlesier *Die Restauration der Götter. Antike Religion und Neo-Paganismus.* Königshausen + Neumann, Würzburg, 26-44.

Finkelstein, Jacob J. (1958) Bible and Babel. A Comparative Study of the Hebrew and Babylonian Religious Spirit, *Commentary* November 1958: 431-444.

Foerster, E. (1902) Babel und Bibel. *Die Christliche Welt* 16, Nr. 8: 188-189.

Herder, Johann Gottfried (1784-91) *Ideen zur Philosophie der Geschichte der Menschheit.* Wiesbaden: R. Löwit; edited by Gerhart Schmidt (no date).

Jensen, Peter (1906) *Das Gilgamesch-Epos in der Weltliteratur.* Strassburg.

Kraus, Hans-Joachim (1969) *Geschichte der historisch-kritischen Erforschung des Alten Testaments.* 2. Auflage. Neukirchener Verlag, Neukirchen-Vluyn.

Kugler, F.X, S.J. (1910) *Im Bannkreis Babels. Panbabylonistische Konstruktionen und Religionsgeschichtliche Tatsachen,* Verlag der Aschendorffschen Buchhandlung, Münster.

Lloyd, Seton (1955) *Foundations in the Dust.* Penguin, Harmondsworth.

Lloyd, Seton (1963) *Mounds of the Near East.* Edinburgh University Press, Edinburgh.

Meade, C. Wade (1974) *Road to Babylon. Development of U.S. Assyriology.* E.J. Brill, Leiden.

Schmidt, P.W. (1908) Panbabylonismus und ethnologischer Elementargedanke. *Mitteilungen der Anthropologischen Gesellschaft in Wien,* XXXVIII: 73-91.

Schrader, E. (1863) *Studien zur Kritik und Erklärung der biblischen Urgeschichte.*

Sprenger, A. (1886) *Babylonien, das reichste Land in der Vorzeit und das lohnendste Kolonisationsfeld für die Gegenwart.* Carl Winter, Heidelberg.

Winckler, Hugo (1901) *Himmels- und Weltenbild der Babylonier als Grundlage der Weltanschauung und Mythologie aller Völker.* Leipzig.

Winckler, Hugo (1905) Die Weltanschauung des Alten Orients, *Ex Oriente Lux* 1: 1-50.

Winckler, Hugo (1906a) Der alte Orient und die Bibel, *Ex Oriente Lux* 2: 1-48.

Winckler, Hugo (1906b) Altorientalische Geschichtsauffassung, *Ex Oriente Lux* 2: 49-112.

Winckler, Hugo (1907) *Die babylonische Geisteskultur in ihren Beziehungen zur Kulturentwicklung der Menschheit.* Quelle & Meyer, Leipzig.

Zimmern, Heinrich (1889) *Die Assyriologie als Hilfswissenschaft für das Studium des Alten Testaments und des klassischen Altertums.*

Note: I thank Professor Jerrold Cooper for valuable suggestions.

The Influence of the Classical Tradition on Anthropology and Exoticism

Fritz W. Kramer

1. The layman, rather than the anthropologist will tend to identify anthropology with exoticism. Only ironically the anthropologist will be ready to refer to his subject matter as being "exotic", although he will admit that parts of it might justifiably be called so. Being an anthropologist, I will reserve the term "exoticism" for that genre in art and literature, which is devoted to portrayals of the exotic world. For the sake of convenience I will use the term, somewhat contrary to English usage, in an extended meaning, in order to include those works of art and literature commonly classified as "primitivism" (cf. Reif 1975: 1-15). It is my intention to show some aspects of a parallel development and some structural relationships between anthropology and exoticism at the turn of the century and during the immediately bordering periods, i.e. from 1860, the founding year of academic anthropology, to 1880 and during the First World War, when Bronislaw Malinowski found his way towards what he called "synthetic ethnography".

2. Let me introduce my main thesis by taking a short-cut, from the field of the visual arts, specifically the art of the nude. Up to the First World War the European arts were unwilling or unable, at least as far as the representation of the exotic nude is concerned, to break away from the classical canon of composition. The classical asymmetry in the logic of the human body, produced by the differentiation of the standing leg from the free leg, was bestowed by the European artist upon people of other cultures (cf. Winkler 1983). This is problematic, since this mode of standing is unique and culture-specific; it occurs regularly in Europe, but only occasionally in other cultures. It is easy for the artist to get his model into an adequate position, if the model is a European, but a man or woman socialized in an other culture has virtually to be forced; and, moreover, the result of this force will always be unsatisfactory in the perspective of the classical canon.

In the end the portrayal will either be entirely untrue or it will show a rather tense figure. The best examples of the former are to be found in the drawings of the artists who accompanied the exploring expeditions of the 18th and early 19th centuries: Although they had actually seen Africans, Indians or Polynesians, their pictures look as if they were copied from an anatomical atlas. Also, examples of rather tense versions of the classical nude are to be found in the exotic photography of around the turn of the century: A sculpture of the Three Graces by Canova, bound to the aesthetic ideal of classicism, may be compared with a photograph, where three African women were forced into the classical positions of the Three Graces (cf. plate 1). The anthropological background of this misleading imposition is evident, when we compare a genuine African mode of standing, as it is paralleled by African art, with its own canon of preferred postures (cf. plate 2). From these two examples we may already conclude that the aesthetics of European art will allow of only inadequate portrayals of other cultures and their people. It was necessary to break away from the classical canon. But this break, which occurred after 1905 in Cubism and Expressionism, characteristically in combination with the discovery of an entirely different aesthetics, the aesthetics of what was then called "primitive art", had furthermore to be applied to the portrayal of the exotic world, as successfully carried out by Emil Nolde and Max Pechstein in 1914, in New Guinea and on Palau. The analysis of these new aesthetics is outside the scope of my present argument, but I might indicate at least one aspect: the synthesis of classical composition and an authentically "exotic" posture. In a recent photograph by Leni Riefenstahl, for instance, the Three Graces reappear, but in the postures of the *nyertun*-dance of the Southeastern Nuba of Kordofan-Province, Sudan (cf. plate 3).

1 a. The Three Graces, by Antonio Canova, Eremitage, Leningrad.
1 b. "Schwarze Grazien im Ringschmuck", in E. Selenka: *Der Schmuck des Menschen*, Berlin 1900, p. 63, fig. 83.

2.a. Art of the Luba-Hemba, coll. C.P. Meulendijk, Rotterdam, reproduced in: R.S. Wassing: *Die Kunst des Schwarzen Afrika*, Stuttgart, Berlin, Köln, Mainz 1977, p. 254, fig. 47.
2.b. "The King with his Favourite Wives", in: C. Egerton: *African Majesty*, London 1938, fig. 26. Detail.

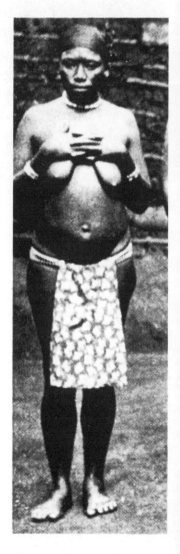

3. Nyertun-Dance, Kao-Nyaro-Fungor, in: Leni Riefenstahl: *Die Nuba von Kau*, München 1982, p. 189.

3. An analogous imprisonment within the confines of classicism is demonstrated by the exotic literature, which flourished at the turn of the century. Since there are some significant divergencies in French, English and German usage, it might be helpful to draw a clear distinction between *realistic* and *exotic* literature. The French adjective *"exotique"* refers to a kind of novel which is set in a locally coloured milieu, i.e. to a novel in the realistic mode. Edmond and Jules de Goncourt, passionate protagonists of realism, for instance, justified their choice of a proletarian milieu in *Germinie Lacerteux* by its analogy to an exotic world: "mais pourquoi ... choisir ces milieux?", they wrote in their *Journal* at December 3rd. 1871:

> Parce que c'est dans le bas que dans l'effacement d'une civilisation se conserve le caractère des choses, des personnes, de la langue, de tout...Pourquoi encore? peut-être parce que je suis un littérateur bien né, et que le peuple, la canaille, si vous voulez, a pour moi l'attrait de populations inconnues, et non découvertes, quelque chose de *l'exotique* que les voyageurs vont chercher... (Auerbach 1982: 463 seq.).

This attitude may be considered as characteristic for realism in general, especially in the 19th century novel, as Erich Auerbach has demonstrated in his study on *Mimesis. Dargestellte Wirklichkeit in der abendländischen Literatur.* Realism is dependent on the possibility of discovering a new social or cultural milieu which has never before been the subject of a novel; its subject-matter is taken from everyday life and treated seriously, thereby mixing ordinariness and decorum and thus breaking away from the classical doctrine of the separation of comedy and tragedy. This is exactly what realism in literature has in common with ethnography or social anthropology. But, strangely enough, the discoveries of milieu in realistic literature never really extended to what in German would be called *das Exotische*, including the exotic and the primitive.

The time for the exotic novel, usually labeled "romance" in English, came, when the reading public was no longer interested in realistic portrayals of the social milieus of modern industrial society, but in an imagination which promises to relieve the strains of everyday existence. The main traits of literary exoticism can be demonstrated in the work of authors like Karl May in Germany and Sir Henry Rider Haggard in England. Both are bound, in a certain way, to the ensemble of classical images; one might point out the relationship of Karl May and Nietzsche, the latter writing his early philosophy, after all, in defence of classicism. But I prefer to take the example of Rider Haggard, since he, but not Karl May, had the advantage of a personal experience with the exotic world of South Africa, where his more famous romances are set (cf. Cohen 1960).

Haggards romance *She* appeared in 1886 and soon became a success, overshadowing that of his widely read predecessor Anthony Trollope, an outstanding protagonist of realism in English literature. For the whole period, which we label "turn of the century", *She* caught hold of the imagination of generations of English readers, among whom anthropologists like Andrew Lang and Bronislaw Malinowski and writers like Kipling and Stevenson, Oscar Wilde and Henry Miller might be mentioned; *She* was interpreted by psychoanalysts and analytical psychologists, and *She* appears even in Freud's *Traumdeutung* (Freud: VI. Kap., Abschn. G VII). At the centre of Haggards strange imagination we find a woman, living for two thousand years in a labyrinth of caves, endowed with eternal youth and esoteric knowledge, ruling a savage tribe, and waiting for her lover, a Greek, who had died two thousand years ago, but is expected to be reborn in the future. Eventually, the young Greek hero is Leo, a beautiful young Englishman, who inherits a mysterious box of documents, among these a map, which leads him into the exotic world of Kôr; he falls into the hands of savages, is rescued and finally enters the realm of a queen called "She-who-must-be-obeyed". She recognizes him as her lost lover and tries to bestow upon him the gift of eternal youth by the "fire of life". But when She enters the burning flame in the midst of a phantastic labyrinth, the fire has not the effect it is expected to have; she suddenly becomes old and dies, and the terrified hero returns to England.

Anthropologists will know the ethnographic background of this weird imagination since the Kriges have published their account of the Lovedu, *The Realm of a Rain Queen* (cf. E. J. and J. D. Krige 1945). Haggard had come to South Africa in the '70s, as secretary of the governor of Natal, and he had a fairly good knowledge of the indigenous population of the Transvaal, where he had served as "Master of the High Courts" from 1877 to 1879. At that time he must have heard rumours of the *modjadji*, queen of the Lovedu, on whom he has written an article as late as 1896 in the *African Review* (Haggard 1896: 639). But Haggard was one of the first modern writers to practice a kind of *écriture automatique*, deliberately without being aware of what he was doing; and he had his own somewhat superstitious theory of his unconscious visions. He thought of his writing as a mediumistic activity, seeing events in far away places he had never visited or in the distant past. He was certain, for instance, that he had described, in one of his novels, the catastrophe of the "Titanic" in all details years before it happened (Haggard 1926: vol. II, 96 seq.). Be that as it may, his description of a South African society in *She* is not his invention, but, in a certain way, ethnographically correct.

The Lovedu are a small, rather peaceful people in Northern Transvaal; although they were not very warlike, they managed to escape the Zulu invasions. The Lovedu inhabit the mountainous, densely forested region, only to be entered by the Duivelskloof, the valley of the devil, which cannot be passed in the rainy season. Caves, terraces, and ruins reminiscent of those of Great Zimbabwe reinforce the mysterious impression, that this topography imposes on foreign visitors. The Lovedu had entered the area around 1600, immigrating from the empire of the *mwene mutapa* and subduing the autochthonous population. They were reigned by kings, but approximately in the year 1800 a woman, called Modjadji, gained access to the throne, and it took not more than 20 years before the neighbouring people were successfully made to believe that she was immortal, since up to the present day she was succeeded in office by her daughter, grand-daughter and great-grand-daughter, all of them by the name of Modjadji, considered as being one person, although in reality dying by ritual suicide. It was firmly believed that the queen killed her enemies by sorcery, that her changing emotions controlled the movements of the clouds, rain and draught, and the spread of locusts; she was said to be immortal and to have four breasts. While she lived in the seclusion of her palace, where she had to give birth to a daughter, whose genitor had to be her own half-brother, her people believed her to sleep by the road like a prostitute. She married, like a king, a large number of women, whom she gave in marriage, without bridal price, to influential men in the provinces, thereby establishing a network of alliances by means of which she was able to rule effectively (cf. E. J. Krige 1975: 55-74; E. J. and J. D. Krige 1945).

It is this African myth of female rule, power, and immortality, which we find at the heart of Haggards romance. Here we have the legend of a primitive, autochthonous population, ruled by a queen endowed with eternal youth and supernatural powers, residing unmarried in the seclusion of her palace; again, we meet the topography of the Duivelskloof, the caves and the ruins, but the African myth is at the same time reshaped so as to fit the demands of rationalistic science on the one hand and humanistic tradition on the other. The secret powers of eternal youth are no longer due to dynastic incest, but to centuries of research into the secrets of nature; the magic of rain and draught has become a metaphor, since her emotions are no longer moving the clouds, but are merely compared to little clouds moving to and fro, with the storm of her will behind them. Now, what is more relevant to my argument, the African *modjadji* is transformed by Haggard into the romantic type of the *"lovely medusa"*, which became, and exactly through the influence of Haggard's *She*, the very epitome of the *femme fatale* so characteristic of *fin-de-siècle* imagination. There is no African prototype for what Haggard reveals as the core of the eternal queen's power, i.e. She is only able to rule her savage subjects by virtue of her deadly looks, killing everybody on whom she turns her eyes in anger... Like the Hadesgorgo of Greek mythology she lives in a subterranean labyrinth of caves and burial chambers. Her face is of a beauty that is too cold and too perfect, capable of horrible distortions; her hair is long and rich, in a rather un-African way, and she wears a "snake-belt", which she winds in her hair, when she enters the fire of life, thus completing the suggestive image of the medusa with the curling snake in her hair. So here we have a perfect embodiment of the "Beauty of the Medusa", a modern, completely aestheticized image of dangerous, fascinating, men-devouring femininity.

4. With this image of the *femme fatale*, modeled at the same time on the prototype of an African rain queen and the ancient myth of medusa, we have already reached the realm of anthropological imagination at the turn of the century. The main historical origin of academic anthropology, as it was established since 1860, was not the enlightened anthropological programme-making of the 18th century, which had no idea of cultural otherness, neither in history nor in contemporary space, but romantic mythology, as founded, already under the influence of historicism, by Herder and Creuzer, whose discoveries were philosophically deepened in Hegel's *Ästhetik* and in Schelling's

Fritz W. Kramer

Philosophie der Mythologie (cf. Kramer 1977). In this tradition Johann Jakob Bachofen published his vision of mother-right. In his first book, *Das Mutterrecht*, 1860, his material was entirely taken from classical sources, especially from those relating to Hellenistic cults and orientalized re-interpretations of Greek mythology. But in his last book, *Antiquarische Briefe*, published in 1880, Bachofen had enlarged his scientific horizon to include ethnographic materials from Africa, Oceania and India; he had also overcome his earlier, rather simplistic ideas on original promiscuity and matriarchy, since he had now discovered the avunculate, i.e. the peculiar relationship between the mother's brother and the sister's son, which since that time has remained a standard item in much of anthropological theorizing (Bachofen1966). Bachofen, who combined his training as a lawyer with his love for mythology and esoteric mysteries, owed his discoveries to his knowledge of Roman Law; and the same source inspired Lewis Henry Morgan and other early anthropologists with a true insight into other types of social organization. But much more than Morgan, to whom the *Antiquarische Briefe* is dedicated, Bachofen was pervaded by the spirit of the classical tradition. The humanistic style of his prose is rhythmically balanced, glowing with metaphors and full of allusions to classical rhetorics. In the history of German Ethnology, if it is allowed to count Bachofen as an ethnologist, as he himself did in his later life, he has no equal in the high literary quality with which he presented his anthropological findings. But he was, of course, equaled by Sir James Frazer, who has been called the "last humanist" by Malinowski, his most faithful and most influential disciple. And what Malinowksi has said in relation to Frazer can also be said in relation to Bachofen:

> He is ... endowed with two great qualities; the artist's power to create a visionary world of his own; and the true scientist's intuitive discrimination between what is relevant and what adventitious, what is fundamental and what secondary. Out of his first virtue came his charm of style; his ability to reshape dull strings of ethnographic evidence into dramatic narrative; his power to create visions of distant land and exotic cultures... (Malinowski 1960: 184).

Those of us who have ever tried to unravel the "dull strings of ethnographic evidence" will, of course, be grateful to anybody capable of transforming empirical facts to readable, even dramatic narrative. But there was a price to be paid for such a visionary transfiguration of exotic cultures and distant lands. There is a sense of the unreal in Bachofen's and Frazer's ethnographic descriptions. If they insisted that their savages must be both incestuous and promiscuous, they failed to convey the sense that what they were writing about existed at all; their readers, just as those of Rider Haggard, were, I suspect, not really interested in the everyday life of the people they were reading about. This is, I think, mainly a result of the classical tradition, under the impact of which they were writing, a tradition shaped according to the demands of dealing with a dead, petrified culture, the classical embodiment of otherness, and, as far as promiscuity and matriarchy are concerned, visions of otherness within another culture. But these cultures of classical antiquity had become an imaginary world, where modern man had become accustomed to seek refuge from the pressures of the present: It was not meant to be real. And the language customarily used in its description was inherited from the antique tradition itself; it was established in the highest decorum, it was voluntarily idealizing, and it was clear that it described an unreal world, or at least a world which was remote and would never and nowhere become an everyday presence. By using modern adaptations of the classical languages, their syntax and rhetorics, the exotic cultures automatically were transformed into an object of infinite longing, without any possible fulfilment.[1]

Let me illustrate the idealizing mode by a brief quotation from Bachofen (1880), which is, moreover, strongly influenced by philosophical idealism:

> Verweilen wir jetzt bei der Betrachtung des Gegensatzes, den die eben entwickelte brahmanische Lehre und das früher entworfene Bild des Familienprinzips der Schlangenvölker darbietet. Dem Tellurismus, der bei den letzteren herrscht, stellt das Priestertum das uranische Lichtprinzip entgegen, dem Prinzipat des gebärenden Muttertums, das auf dem Vorbild der Erde beruht, jenen des zeugenden Vatertums, das auf die

---

[1] It may well be that Frazer was much more rooted in the tradition of 18th century enlightenment, while Bachofen was an extreme protagonist of romantic mythology and counter-enlightenment. But a close parallel between Bachofen and Frazer in relation to the classical tradition is at least brought out by a more 'continental' reading of Frazer. That such a 'continental' reading may well be relevant is demonstrated by the case of Malinowskis interpretation of *The Golden Bough*.

himmlische Lichtpotenz zurückgeht. Mit diesem wird die Idee einer Auseinanderfolge der successiven Geschlechter zuerst in die Welt eingeführt. Dem tellurischen Mutterprinzip fehlt sie ganz. Hier herrscht statt des Auseinander von Vater und Sohn das Nebeneinander von Bruder und Schwester, statt der Einheit in der Vielzahl der sich folgenden Söhne und Sohnessöhne die Verschiedenheit der gleich den Blättern der Bäume in ewiger Gleichförmigkeit sich reproduzierenden und wieder verschwindenden individuellen Existenzen (Bachofen 1966: 68 seq.).

It was Malinowski who finally discovered the sound sociological core of Bachofen's observation: the emphasis on the relationship between brother and sister in matriliny; but in Bachofen's vision there was much more than that: "gleich den Blättern der Bäume" is a classical reminiscence of the vanity of life in matriarchy, idealized and positively evaluated like the eternal return in Nietzsche's coeval philosophy.

It has often been said that the unreal and visionary colouring of the writings of early anthropologists, like Frazer and Bachofen, was due to the fact, that they were "armchair-anthropologists" and never went to the field. Therefore I would like to point to such a towering figure as Leo Frobenius who had many years of first hand experience in African cultures, and yet did not break with the classical tradition, relabelling Zimbabwe as "Erythräa" or African agriculturalists, quoting Herodotus, as "die unfehlbaren Äthiopen" (Kramer 1986: 258-70). It is evident, I think, that the dichotomy is not between armchair-anthropologist and field-worker, but between the impact of the classical tradition on the one hand, and an empirical and realistic mode on the other. It is time now to have a look at the alternative.

5. The emphasis on a scientific approach, so characteristic for the second half of the 19th century, was essentially a means of getting away from the restrictions of the classical tradition, of breaking up the limitations of its forms, of enlarging the range of subject matters, and of escaping the necessities of idealization. Even a classically minded scholar like Bachofen found it useful to justify his choice of such unclassical subjects as primitive promiscuity and matriarchy by stressing the scientist's obligation towards a truthful interpretation of his authentic sources, even if

these were immoral or ugly.[2] While the artist was still put under the obligation to choose sublime and solemn subjects for his creations, the scientist was already free to choose any subject which was of interest to him. So the emphasis on a scientific approach was especially important in early anthropology, since anthropology was not concerned with the admired and worthy cultures of classical antiquity, not even with the literate cultures of China, India or the Near East, but with the humble beginnings of mankind, the primitive, naked savage, living in heathen darkness, full of superstitious fears and anxieties, bound to immoral customs like cannibalism and promiscuity.

To give more plasticity to my argument, it may be useful to go back once more to the justification Edmond and Jules de Goncourt had given for their choice of a proletarian milieu in *Germine Lacerteux*. In their preface to this novel they wrote:

> Aujourd'hui que le Roman s'élargit et grandit, qu'il commence à être la grande forme sérieuse, passionné, vivante de l'étude littéraire et de l'enquête sociale, qu'il devient, par l'analyse et par la recherche psychologique, l'Histoire morale contemporaine; aujourd'hui que le Roman s'est imposé les études et les devoirs de la science, il peut en revendiquer les libertés et les franchises (Auerbach 1982: 460 seq.).

If even a littérateur and aesthete could justify his novel by giving emphasis to the scientific research behind it, by introducing concepts like *récherche, analyse* and even *enquête*, we should, of course, not be surprised to find the same emphasis in the anthropology of the time, which was radically positivistic, at least in its programme, not in spite of its exoticism, but exactly because it had to justify its exotic subject matter. On the other hand, the quotation from the Goncourts demonstrates clearly that a scientific enquête was not seen as being in any contradiction with the new aesthetics of literary realism, it was only a negation of the classical doctrine. Therefore there remained room enough for the arts in anthropology, in spite of its positivistic appearance.

---

[2] "Das Zeugnis der Geschichte verbietet, den Einflüsterungen des Stolzes und der Eigenliebe Gehör zu geben und den äusserst langsamen Fortschritt der Menschheit zu ehelicher Gesittung in Zweifel zu ziehen. Mit erdrückendem Gewichte dringt die Phalanx völlig historischer Nachrichten auf uns ein und macht jeden Widerstand, jede Verteidigung unmöglich" (Bachofen 1948: 36 seq.).

Fritz W. Kramer

The liberation from the restrictions of the classical tradition the Goncourts had achieved was, of course, much less radical as it was felt to be in the 1860ies. And the same may be said in relation to anthropology. Take, for instance, the pioneering work of Adolf Bastian, who is generally credited as the founding father of anthropology in Germany. Bastian, I suppose, was the best representative of those anthropologists who were trying to force the philological model developed in classical studies on the study of cultures that were essentially oral cultures. The restrictions bestowed upon the arts by the classical canon, upon literature by the obligations of decorum, and upon anthropology, as practiced by Frazer, Bachofen and others, return within this framework as limitations of the philological method, which may be inadequate when applied to oral traditions. Since there were no texts to be studied, philological substitutes had to be produced. For Bastian these substitutes were the artifacts of material culture, which could be collected in museums. "Betreffs der Culturvölker," he argued, referring to the so-called high cultures with a literary tradition,

> gelten in der Philologie diejenigen Gesichtspunkte, wie grammatisch ausverfolgt, zum Einblick in das sprachlich innere Walten, dessen Verkörperungen in der Schrift sich niedergelegt finden... In der Linguistik, bei schriftlosen Wildstämmen, stehen an der Stelle der Texte (und Handschriften) die ergologischen Sammlungen, um unter hermeneutischer Entzifferung (Symbolisch allegorisierender Zeichen) das Gedankenleben aus dem herauszulesen, woran, als Manifestem, mit der Hand gemacht, dasjenige kleben geblieben ist, was (psychisch) von dem Handwerker (oder Künstler) hineingelegt wurde (Bastian: 50).

The disintegration of grammar and style in Bastian's writing is evident; we are far away from Bachofen's lucidity and cunning rhetorics or Frazer's gloomy prose with its purple and gold, green and blue; and yet even in Bastian's rather tense syntax we can discover the formative influence of the classical languages; and this is in line with his argument, the substitution of classical texts by primitive artifacts. This situation did not change until World War I, since Franz Boas and his school, as well as German missionaries like Hofmayr and Westermann, Gutmann and Spieth, were still mainly concerned with the collection of texts, now taken from the oral tradition and written down by them in the vernacular. The humanistic model still remained the main prototype of this kind of anthropology, although it would have been possible to communicate with the people under study without the medium of written texts.

Again the shortcomings of the classical model were by no means due to any lack of fieldwork, as has been repeated again and again by historians of anthropology, especially in Great Britain. Boas' fieldwork among the Kwakiutl is well known, but missionaries with considerable anthropological skill like Spieth or Hofmayr stayed in the field for many years and had a command of the vernacular, which, I think, no modern professional anthropologist would dare to claim. And yet as far as Bastian is concerned, it should be enough to recollect that he had worked in the exotic field for 24 years.

One might rather say that he had seen too much, visiting practically all parts of the globe. In the end he was obsessed by the idea that all human cultures were built on the same foundation, on what he called the "Elementargedanken":

> Es fand sich leicht genug die Maskierung abgezogen, die Auftünchung fortgewaschen, und dann, siehe da! zu Boden unterst lag Er überall offenkundig vor Augen: der ärmlich-arme Elementargedanke nackt und bloss, ein und derselbe überall, in allen fünf Continenten derselbe (unter topischen Varianten), - so, wie ihn der Wildstand gedacht, auch unter den cerebralen Kartenhäusern der Civilisation verborgen (Bastian 1900: 13).

It is impossible to enlist exactly what Bastian counted as "Elementargedanken", since he was rather "muddle-headed", as Paul Radin has said (Radin 1965: 72). But it is clear that the concept itself was derived from Friedrich Creuzer, a classical philologist, who in his turn had but reformulated the neoplatonic concept of ideas (Kramer 1977: 80).

If fieldwork was not in itself a guarantee for the liberation from the impact of the classical model, it gave at least a chance to discover the inappropriateness of philology for the description of an oral culture. And this was certainly the only chance there was. One of the few among the early anthropological fieldworkers who came to recognize the advantages of direct communication with the people of another culture, the advantages of what was later to be called "participant observation", completely absent in the classical model, was Johann Stanislaus Kubary (cf. Probst 1983:

23-56). Kubary, whose work is today known only to a few specialists, was the true pioneer of "participant observation", as Malinowski following his lead has acknowledged in his *Diary*. Kubary's ethnography happily helps me to return to the subject of female rule, commonly called "matriarchy" by the turn of the century, which we have already observed in the versions of Rider Haggard's *She* and Bachofen's *Mutterrecht*. At the same time Kubary's ethnography will be a good example of a kind of anthropological writing, which has freed itself of the idealistic strains of the classical tradition, although without having found a new form of literary or scientific representation and portrayal.

Kubary had a touch of the great adventurer, reminiscent of some characters in the novels of Joseph Conrad, of Caspar Almayer in *Almayer's Folly* and, in some aspects, even of Mr. Kurtz in *Heart of Darkness*. Born as the son of a German mother and a Hungarian father, he grew up in Poland, became involved in a political conspiracy and escaped from being arrested by emigrating to Germany, where he was engaged by the ethnographic museum Godeffroy in Hamburg to collect ethnographic specimen in the South Seas. In the 1870s and '80s he carried out field research in Micronesia, mainly on the Caroline Islands. He was most successful on Palau, where he resided for many years, learning the vernacular, until he was finally accepted by the natives as *rupak*, speaker and chief; he was even appointed to the office of an official representative of the Palauans in their dealings with the colonial administration. He married a native woman of Ponape, Yelirt, and consequently he was ostracized by the white residents. His profound ethnographic writings were published in Germany, but disapproved by his colleagues. Thus, he had no means of earning his living; he had to work for a plantation, and in this job he developed an extreme unscrupulousness and cruelty towards the natives. He started to drink and finally, in 1896, ended his life by committing suicide.

Kubary was accused by contemporary anthropologists of having "gone native", and, what is more to the point of my argument, of giving ethnographic descriptions which appeared to be too complex. Rudolf Virchow, for instance, wrote in his review of Kubary's best book *Die socialen Einrichtungen der Pelauer*, published in 1885:

> In der That sind die gesammten gesellschaftlichen Stammes-Gebräuche der Pelauer von so verwickelter Natur..., daß nur

ein auf lange Erfahrung gestützter Beobachter ein so einge-
hendes Bild davon entwerfen konnte (Virchow 1885: 204).

Since Virchow's review is entirely slating, this was clearly a
reproach. Kubary had written a lucid and meticulous account of a
matrilineal society, actually the first which was really up to our
present and quite advanced standards in anthropology. The sad
fact is that at the turn of the century nobody, not even the
anthropologist, was interested in the social organization of
matriliny. Characteristically, Bastian did not even mention the
concrete results of Kubary's research in his preface to *Sociale
Einrichtungen*; he preferred to speculate on some superficial
similarities between Kubary's findings and the then current
rather lurid phantasies of promiscuity and gynaecocracy, marriage by capture of females or the origin of the state. That was the
kind of stuff the European reader at the turn of the century
wanted to read; the emphasis on a strictly scientific approach was
merely a facade, a farce and a disguise for the repressed phantasies
of the bourgeois. It is sad to read Kubary, in a letter to Bastian,
accusing himself of having lived in the "folly" that his research
might be of scientific interest:

> In Hinsicht auf die Mittheilung, dass meine Manuskripte zwar
> gründlich, mir aber praktisch nicht nützlich sind, so kann ich
> nicht helfen, ein ausdrückliches Bedauern auszudrücken, da
> ich dieselben in Zeiten physischer Noth ausarbeitete und in
> dem Wahne befangen war, dass sie bei uns erwünscht sein
> dürften (Probst 1983: 48).

There is only one real excuse for Kubary's contemporaries, i.e.
that they were simply unable to understand what he was writing
about. Kubary had not yet mastered the problem of a readable
representation of ethnographic facts. I quote a very short example
of what is meant here:

> In einem Falle sah ich den Aybadul von Korryor einem
> Kaldebekel Strafgeld zahlen, weil einer seiner ngaleki unter
> Bul Kabuy pflückte (Probst 1983: 49)

The literary form of ethnography could not have been taken
from classical models, as I hope I have demonstrated. But, on the
other hand, a high scientific standard did not suffice to convey a
concrete and vivid idea of what had been observed. Where

Fritz W. Kramer

Bachofen and Frazer were too much of an artist, Kubary was too much of a scientist.

6. It was Bronislaw Malinowski who found a unique and admirable synthesis between the exotistic tradition of humanism as represented by Frazer and the scientific participant observation of the exotic world practised by Kubary. On the basis of his Trobriand fieldwork he published, already in 1915, his long essay on *Baloma. The Spirits of the Dead in the Trobriand Islands*, containing in a nut-shell the whole of what he called his "new humanism" or, alternatively, his "synthetic ethnography". The leading idea behind his essay was to combine an exotic colouring with sound ethnographic observations. In order to achieve this end, he simply told of the exotic imagination of the Trobrianders themselves, their phantasies of an island paradise, where the dead would live in eternal youth and everlasting erotic fulfilment. Here were all the elements of Bachofen's or Frazer's visions of a "matriarchal" society. The enchantment of an erotic paradise was conveyed to the reader, but at the same time the very quality of the material, its imaginary character, led additional emphasis to the soundness of the ethnographic description, since it was obvious now that a people which develops this kind of imagination could not actually live in a state similar to its own dreams. The ethnographic reality of everyday life *had* to be different. A similar point may be made in relation to Malinowski's first monograph on the Trobrianders, bearing the title *Argonauts of the Western Pacific*. Characteristically, the title can be read in the framework of the classical tradition, as referring not to the savages, but to the *Argonauts* of the Western Pacific; and it may as well be read as a break with that tradition, since it also means that the book is not on the *classical* Argonauts but on those of the *Western Pacific*. It is tempting to go into the details of Malinowski's composition, which satisfies both scientific and literary demands, the canon of the humanistic tradition and the standards of scientific research and authenticity (cf. Kramer 1979: 558-70). But, I think, the basic lines are already apparent. To find this solution, Malinowski could make use of the foundations laid by Frazer, Kubary and others; but a basic condition for his success in doing so was that the reading public was now ready to appreciate cultural otherness, mainly, I think, because European selfconsciousness had been shattered by the cultural crisis following the war. The reader was now less ethnocentric than he had been at

the turn of the century, when he had a longing for mythical images of otherness, but was not interested in a reality completely different from his own.

7. Let me summarize my argument. At the turn of the century no anthropologist seems to have been aware of having to choose between the arts and the sciences; in the humanistic tradition authors like Bachofen and Frazer had a perfect command of the traditional means of "artistic" writing about cultural otherness; but these stylistic means were inadequate for the description of oral cultures. They had to justify the choice of their subject matter by claiming the liberty of the scientist, because otherwise to them it would not have appeared to be legitimate to write about seemingly immoral manners and customs. If they claimed to have a scientific approach, they did not essentially differ from writers of realistic novels claiming the same basis in the same demand for justifications of "exoticism" in the choice of social milieus. Those who actually developed a sound scientific approach, like Kubary, were reproached by their colleagues. Just like the exotistic artist at the turn of the century aimed at an alienation of the classical canon, using familiar forms in an unfamiliar setting, the anthropologist, in spite of his radically positivistic appearance, was, in fact, formed by the humanistic tradition; and in spite of his exotic subject matter he was defending the classical canon of humanism. The firm belief in philosophical idealism and in the universal adequacy of the classical forms was not destroyed before the War. And this belief was the basis of Nietzsche's programme: "die Wissenschaft unter der Optik der Kunst sehen, die Kunst aber unter der Optik des Lebens."

Fritz W. Kramer

Bibliography

Auerbach, E. (1982) *Mimesis; dargestellte Wirklichkeit in der abendländischhen Literatur* (1946), Francke, Bern.

Bachofen, J. J. (1948) *Das Mutterrecht* (= *Gesammelte Werke*, ed. by Karl Meuli, vol. 2), Schwabe und Co., Basel and Stuttgart.

Bachofen, J. J. (1966) *Antiquarische Briefe* (= *Johann Jakob Bachofens Gesammelte Werke*, ed. by Johannes Dörmann and Walter Strasser, vol 8), Schwabe und Co., Basel and Stuttgart.

Bastian, A. (n.d.) Randglossen zur Erörterung schwebender Fragen in der Mensch- und Völkerkunde, *Ethnologisches Notizblatt* II, 1, Beilage.

Bastian, A. (1900) *Die Völkerkunde und der Völkerverkehr unter seiner Rückwirkung auf die Volksgeschichte*, Berlin.

Cohen, M. N. (1960) *Rider Haggard; His Life and Works*, London.

Freud, S. *Traumdeutung*, VI Kap., Abschn. G VII.

Haggard, H. R. (1896) The Death of Majajie, *African Review*, VIII, Sept. 19.

Haggard, H. R. (1926) *The Days of my Life*, London, vol. II.

Kramer, F. W. (1977) *Verkehrte Welten; zur imaginären Ethnographie des 19. Jahrhunderts*, Syndikat, Frankfurt a. M.

Kramer, F. W. (1979) Nachwort, in: Bronislaw Malinowski: *Argonauten des Westlichen Pazifik*, ed. by F. Kramer, Syndikat, Frankfurt a. M., 558-70.

Kramer, F. W. (1986) Die Aktualität des Exotischen; der Fall der "Kulturmorphologie" von Frobenius und Jensen, in: *Die Restauration der Götter; Antike Religion und Neo-Paganismus*, ed. by Richard Faber and Renate Schlesier, Königshausen and Neumann, Würzburg, 258-70.

Krige, E. J. (1975) Divine Kingship, Change and Development, in: *Studies in African Social Anthropology*, ed. by Meyer Fortes and Sheila Patterson, Academic Press, London, New York, San Francisco, 55-74.

Krige, E. J. and Krige, J. D. (1945) *The Realm of a Rain Queen*. Oxford University Press, London.

Malinowski, B. (1960) Sir James Frazer; a Biographical Appreciation, in: Bronislaw Malinowski: *A Scientific Theory of Culture and Other Essays* (1944), Oxford University Press, New York.

Probst, P. (1983) Beobachtung und Methode; Johann Stanislaus Kubary als Reisender und Ethnograph im Spiegel seiner Briefe an Adolf Bastian, in: *Baessler-Archiv*, N. F. 31, 23-56.

Radin, P. (1965)*The Method and Theory of Ethnology; an Essay in Criticism*. (1933), Basic Books, New York, London.

Reif, W. (1975) *Zivilisationsflucht und literarische Wunschräume; der exotistische Roman im ersten Viertel des 20. Jahrhunderts*, Metzler, Stuttgart.

Winkler, D. (1983) *Die Pose im exotischen Bild*, Unpubl. MA-Dissertation, Freie Universität Berlin.

Virchow, R. (1885) Rezension von Kubarys "Die socialen Einrichtungen der Pelauer", in: *Zeitschrift für Ethnologie* 17.

European Barbarism and the Search for Authenticity

Mike Rowlands

> What can oppose the decline of the West is not a resurrected culture but the utopia that is silently contained in the image of its decline (Adorno 1967: 72)

The major theme of this paper is captured in this quote from Adorno. It accurately conveys the feeling that the rich humanistic discourse that grew up in Europe on cultural difference and unity is in fact rooted in a peculiarly baleful view of its own past and driving force in history. This I take to be codified in a consistent reference to barbarism in European definitions of its own relationship to the civilized. The paradox this presents is that barbarism is presented as both an energising force that has provided European history (and some of its prehistory), with a characteristic vigour and capacity for growth, and as a destructive and corrosive evil that erodes culture, rootedness, unity and a sense of belonging.

Barbarism has been identified with historical events such as the Germanic migrations, or with the alienation of capitalist commodity production. Yet what these identifications share is a consistent theme of heterogeneity and fragmentation - "civilization as the most external and artificial state of which humanity is capable" (Spengler 1926: 356) opposed to culture as the correspondence of the person with a sense of place, with an unaltered and continuous past and an identity with nature and the universe. If this represented simply a dualism maintaining a precarious balance in legitimacy, I suppose this paper would be another disquisition on the false coinage that societies are capable of paying themselves in order to repetitively mis-recognize (méconnaissance) the full nature of their own social practices. My argument will deny this and claim instead that the European discourse on barbarism holds out a continuous promise of expanding an imperfect humanity within a logic that claims that this can only be achieved through harnessing to it the forces which bring about its own destruction and transformation.

The role of anthropology in this civilizing mission has been well defined by Lévi-Strauss in his *Place of Anthropology in the Social Sciences* where he claimed "that anthropology's most important contribution to the social sciences is to have introduced, if unknowingly, this fundamental distinction between two types of social existence: a way of life recognized at the outset as traditional and archaic and characterized as "authentic societies" and a more modern form of existence, from which the first named type is not absent but where groups that are not completely, or are imperfectly "authentic" are organized within a much larger and specifically "unauthentic" system" (Lévi-Strauss 1963: 367). Here in contrast to the usual modern/traditional dichotomy, Lévi-Strauss holds out some hope that anthropology can expand the imperfectly authentic within the iron cage of modernity.

Sir Keith Joseph, the right wing former Minister of Education in England, recently advised British anthropology that it possessed a humanizing mission to exert a civilizing influence in modern, pluralistic societies. A role that I heard recalled recently at a meeting of Heads of Departments of Anthropology in Britain as the argument anthropology should use to fend off further government cuts in the funding of overseas research. Such anecdotes are part of a common academic experience these days, but such views would be unutterably depressing unless I could take heart and assure myself that this is historically conditioned, a product of an epoch, and not one of those cultural invariants of Braudelian proportions. The problem which confronts us is well described by Walter Benjamin's private appraisal of Ernst Bloch's book *Heritage of our Times* :

> The serious objection which I have of this book is that it in absolutely no way corresponds to the conditions in which it appears, but rather takes its place inappropriately, like a great lord, who arriving at the scene of an area devastated by an earthquake can find nothing more urgent to do than to spread out the Persian carpets - which by the way are already somewhat moth-eaten, and to display the somewhat tarnished golden and silver vessels, and the already faded brocade and damask garments which his servants had brought (Benjamin 1966, Brief II pp 648 and 649.)

This then is the ideology with which this paper is engaged taking the intellectually organizing practices of Barbarism as a phenomenon of the 20th century as my central theme.

Barbarism and European Authenticity

Ironically, Ernst Bloch's book is particularly concerned with identifying those ideological remnants of past epochs that the contemporary world appropriates (in his case fascism) to gain a place in a powerful, if fragmentary, anti-capitalist heritage.

Undoubtedly one of these fragments was the slow but certain resuscitation in humanism of an authentic barbarian Europe. Since the Renaissance, European barbarism had been held responsible for the destruction of the civilized world from which Europe was only gradually recovering. To the humanists of the Renaissance everything that lay between them and their admired classical models smacked of Northern barbarism; an attitude harshly expressed by Filaretes in 1450:

> Cursed be the man who invented this cursed Gothic architecture; only a barbarian people could have brought it to Italy (quoted in Dopsch 1937: 2).

To them the fall of the Roman Empire was a calamity and the Barbarian north, and in particular the Germans were responsible for it. Northern barbarism unable to rise above pillage and rape, had destroyed the most perfect, divine creation and more significantly, had broken the transfer of civilized identity from the classical to the modern world. The French Enlightenment adopted a broadly similar Romanophile attitude; Voltaire called the Greeks and the Romans the two most gifted nations of the world. Montesquieu gave a detailed description of the peculiar Roman talent for trade ,and identified as a sign of Barbarism the tendency to see trade as a mere object of plunder and towns as things to be sacked. Voltaire's picture of the transition from the Roman to the German world is particularly illuminating since it privileges the modern virtue of unified language as the defining feature of a civilized existence:

> If we pass from the history of the Roman empire to that of the peoples who destroyed it in the West, we feel like travellers who leave a splendid city to find themselves in a thorny waste. Twenty barbaric dialects are the heirs of the beautiful Latin

language, which was spoken from Illyria to the Atlas Mountains (Essai sur les moeurs; end of chap. 12).

Europe's shameful barbarism - its guilt - issues from its compulsion for fragmenting and heterogeneous forms of existence: like a dead hand on all it touches, it forms a monstrous negation of a civilization of unity and harmony bequeathed to it or encountered by it from the outside.

Of course those held most responsible - to whom the accusing finger points - are German. If the Renaissance and French Enlightenment recaptured the civilizational ethos, then the concept of Central Europe captures the other side of the paradox (Dopsch 1937: 6). The German tribes living a pastoral life amidst forests, marshes and lakes had escaped the tyranny of industry and town life. Moreover, they had developed personal freedom and democracy to a high degree; the Germanic mode of production based on individual ownership of land, encouraged the absence of authority in the German mark association and the participation of all free men in government. The German national history was justified by the contribution that tribal life had made to the creation of the Middle Ages as a unique and balanced synthesis of Romano-Germanic elements and their determination by a system of agriculture adapted to natural conditions. But German history rends itself on the unfortunate implication that this also denies possession of the state as a natural condition. German barbarians are historically free but the very conditions of that authentic freedom in culture denies them a conception of the State and access to modernity.

The ideal German state should instead embrace the organic identity of the homestead. In the later 19th century this was to be transformed into the 'Völkish' idea of the superiority of the peasantry as closer to nature in contrast to a mechanical and materialist civilization (Mosse 1966: 23). One of the consequences of the re-evaluation of primitive barbarism is that a former shameful tribal stage, destroyer of civilization, now becomes the universal peasant communal system of the Russian Mir; the Slavic Zadruga; the Indian village of the Asiatic mode of production or the African farmer. From Sumner Maine's Ancient Law in 1861 to Morgan's Ancient Society in 1877, more than seven major works were published in German and several more in France, by e.g. Viollet, Geoffroy and Laveleye, on the theme that modern European societies were the result of a continuous progress from communal land-owning traditional societies to modern private individualism. The primitive communal conditions of the barbarians were to be found everywhere in world history and had evolved to a level consistent with Roman juridical ownership at the time of great Migrations. Fustel de Coulanges, Richard Hildebrand and Ernst Mayer would all repeatedly attack the communistic idea of Early European origins but to no avail.

Primitive collectivism constructed as a universal form of traditional society allowed at a crucial period a common and historically universal basis for nation state formation in Europe (and elsewhere as well (cf. Anderson 1983). This is accompanied by the gradual domestication of European barbarism as the historical destroyers of civilization. In 1872, Gaston Paris a linguistic scholar of great repute, had proved that German barbarian tribes had not been the assailants or destroyers of classical civilization but merely strangers to it. They came with cultural difference, settled and fused with Gallo-Romano elements but continued a process of interaction already initiated in the later Roman Empire. What was revived was the emphasis on cultural synthesis based on respect for cultural difference - in custom, education and culture - rather than conquest and unity established on violence and destruction which was now credited with being a malignant invention of the Renaissance.

As a consequence of this return to virtue, a barbarian prehistory of Europe gains respect, is considered worthy of investigation and undergoes a major development in the period from 1870 to 1895. In 1872 De Mortillet publishes his sub-divisions of the Stone Age. In 1875 Hildebrand produces the Halstatt and La Tene divisions of the European Iron Age and then years later Tischler sub-divides the La Tene Iron Age of central Europe (contemporary with the expansion of Rome) into three sub-phases. By 1895 Montelius publishes his five stage divisions of the Bronze age of Northern Europe and ten years later his four stage sequence for the Italian Bronze Age. Reinecke uses this scheme later as a basis for working out the Bronze Age chronology for Central Europe. All of this chronologizing is done on the strictest 'scientific basis' of technological-typological development whilst social and economic speculations on the meaning of the different assemblages are discouraged. In 1911, Kossina published his general account of the prehistory of Northern Europe, including a fully developed 'racist account' of Aryan origins based on the assumptions that the culture areas defined by Montelius on the basis of artefact comparisons and distributions, now unquestionably represent peoples. Moreover, these 'peoples', by sharing the same material culture, must have spoken the same language and be the product of a common history. The Nordic area, Kossina claimed, had been occupied by the same people for millennia, which later would be used to justify militaristic expansion to reclaim areas shown by archaeology to have been originally German or non-Slavic. Moreover, it is in this relatively short period that Schlieman's excavations at Troy, Tiryns and Mycenae; Arthur Evan's at Knossos; Petrie's campaigns on the pre-dynastic periods in Egypt develop a pre-classical or prehistoric antiquity for the Classical World to match that of a Barbarian Europe. The borrowings and interdependence of the classical world and Europe can now be extended back into the remoter prehistories of both areas.

Within these various recontextualizations of the meaning of barbarism there exists the sub-text that destruction and violence can be justified as creative acts by the European myth of origin. In the *Persian Letters*, Montesquieu had depicted the cosmic role of the Oriental Despot as one of bringing a chaotic periphery into order. The Asiatic King brings fertility, towns grow and irrigation networks flow as a result of cosmic incorporation - imaged as a royal hunt in the wild periphery. He contrasts this to European civilization as the bringer of disorder to a perfect if decaying order. Living in the fear that their acts will correspond to this stereotype, European myths are structured on the transcendent principle that massacres, as in the Song of Roland, should be turned into the saving of Christendom; looting and plundering should be disguised, as in the Chivalric tradition, to become acts worthy of the defence of a united Christendom. To punish and to eradicate the unworthy is also the object of the purification of violence. It is an honourable exercise in discriminate violence to drive Arabs out of Spain, to torture heretics or to conquer Pagan Irish who curiously never learn their lesson and need to be continuously reconquered to keep their barbarism in check. The naturalizing of barbarism and its conversion from pillage, violation and destruction to a creative force - freeing and liberating and surgically civilizing betrays only too clearly a humility which is a kind of inverted hubris. To quote from the conclusion of a paper by Jacqueline Kaye that has influenced me here "John Wycliffe wrote that the leading characteristics of Islam and the

Western church were the same: pride, cupidity, the desire for power, lust for possessions, violence and the preference for human ingenuity over the work of God" (Kaye 1985: 70).

Barbarism and Modernity

Myths of origin have the power to continuously recapitulate their conditions of existence.

In the period from 1890 to after the First World War, there developed in Europe and particularly in France and Germany, fears of moral decadence and beliefs in the need for institutional renewal and reintegration of the masses through new forms of leadership. Gustave Le Bon expressed a theory of leadership, reflected in the growth of such cults and faiths when he recognized that the leader himself was hypnotized by the beliefs of which he had become an apostle (Le Bon 1896).

There are two themes that in particular influenced the writers on mass society in this period. One that preoccupies most of them is the fear of a world in moral decline and the kind of action required to regenerate an epic state of mind. And the second which grew apace during this period but in particular during and after the First World War, was the critique of violence and the advocacy of a humanism that could be genuinely non-violent. Both proposed different anthropologies which were taken up in different national cultures in different ways. And of course to a large extent the options are still very much with us now but in potentially even more catastrophic circumstances.

A pre-requisite for the first of these had been the growth of pragmatism as fin-de-siècle empiricists discovered 'unreason' with the most rational scientific procedures. The acceptance of social engineering leads to a view of society as an object to be worked upon and reshaped. We have entered the era, claimed Le Bon, where "the unconscious action of crowds, substituting itself for the conscious activity of individuals, is one of the principal characteristics of the present age" (Le Bon 1895: i). More significant is the idea attributed to Sorel, that crowd life is a surer portrait of natural man and a better indicator of authentic human needs than any other form of human behaviour (Nye 1973). In contrast to the false rhetoric of parliamentary democracies where men are dressed in their public masks, crowd life reveals the conditions necessary for the return to political ideals and the regeneration of moral being. And Sorel's theory of social myths

Mike   Rowlands

that would rouse the proletariat to class struggle or Le Bon's recognition of the suggestive power of images and words to speed up social transformation, continues a belief in the contribution of social science to practical politics that many of their contemporaries shared and which right wing working class movements and the fascisms were later to fully exploit (Sorel 1950: 259-264).

But the form of social sclerosis most feared by Sorel was the widely held belief that the bourgeoisie had missed their historical destiny and were now incapable of maintaining dynamic social progress. A timorous, humanitarian capitalist class had lost the conquering, insatiable and pitiless spirit on which European economic progress had depended so far. The same argument is found in Max Weber's condemnation of the German bourgeoisie for eluding their historical destiny by establishing an alliance with the Junker aristocracy to preserve the status quo (Weber quoted in Beetham 1974:157).

A parliamentary socialism had also defanged a revolutionary working class whose demands were met by the very threat of disorder and strike action. The Dreyfus affair had for Sorel and others shown the depth of moral decadence that France could be driven to. Whilst Durkheim and his school attempted to elaborate a theory of social cohesion in order to save democratic institutions in France, collective psychologists exhibited a political bias that was anti-democratic, elitist and pessimistic and denied any rationality to the actions of mass movements (Nye 1973: 429).

The rugged individualism of the capitalist entrepreneurial spirit had somehow been evaded and decadence and decline were preconstituted as inevitable consequences by the European myths of cyclical histories. In drawing upon Vico's philosophy of history, Sorel made great play of what would happen if there was a revolution in a period when capitalism was still dominant but degenerate. The result would be the equivalent of what had happened with the fall of the Roman Empire "there had to be almost four centuries of barbarism before a progressive movement showed itself: society was compelled to descend to a state not far removed from its origins".

Civilized socialism, as he contemptuously referred to parliamentary socialism, rather than saving civilization would remove the dynamic of struggle and progress, would bring decadence and a return to barbarism. The unselfish, heroic and disciplined use of violence was the only means of reopening the class struggle so that human action could be put back upon authentic lines, thus ensuring, as he put it, that "those whose benevolence would protect the workers would be repaid with black ingratitude" (Sorel 1950).

And it is a special type of violence that is needed - discriminate and socially healthy cruelty is advocated - "It is not perhaps the most appropriate method of obtaining immediate material advantages," Sorel wrote, "but it may save the world from Barbarism" (Sorel 1950).

Barbarism, force and conspicuous consumption

One of the instincts that incites revolts against positivism claims that the individual need not be ground down by iron laws of social and technological development. Hence *enragés* like Sorel and Le Bon are extreme but still part of the spirit of the age. However, whilst Weber might rage against bureaucracy as the iron cage or Simmel against the increasing separation of objective from subjective culture, we read them as great social thinkers precisely because they are bracketed as still concerned with advocating the possibility of moral forms of sociation in what was seen as a new and horrific age.

Barbarian thinkers such as Sorel, Veblen, Spengler and Bataille have acquired unsavoury reputations precisely because they put such thoughts in doubt and did not appear to give much for the forms of sociation on offer when they considered the costs involved for the individual. It is quite appropriate for example that Veblen should be best remembered for the notion of conspicuous consumption and Bataille for writing a doctorate on waste. That Veblen should have ended his life living in a log cabin in the Rockies and also be the writer who points out to us that all the elements of modernity that had been optimistically pointed out as indicating what humans were capable of when freed from the realm of necessity, were in fact a regression to traditional archaic practices. The dawning horror was nothing but a display of power, loot and profit in modern barbarian culture. All sports for instance, Veblen argued, are outbursts of violence, oppression and the predatory spirit and include not merely the drive to do violence to others but also to suffer cruelty to oneself. No doubt such exhibitions of ferocity and cunning could be taken, as Adorno later suggested, to be models for totalitarian mass rallies (Adorno 1967).

Veblen's talk of the barbarian normal was also to suggest to Adorno the thesis that it was steadily reproduced in direct proportion to human technical control. Barbarism is equated by him with a 19th century view of German traditional culture and thus with a natural state, where personal and collective ties dominate with all their oppression and unfreedom for the individual. However, this is not the barbarism of other contemporary thinkers. They celebrate instead the ruthless pursuit of profit and entrepreneurial skill, debate the principles of waste, futility and ferocity, whilst at the same time indulging in nostalgic moments of traditional reverie. But it is a reverie which in the end extends the same principles into a traditional past in order to show *'plus ça change, plus c'est la meme chose'*.

"Veblen invents the instinct of workmanship only incidentally in order to bring paradise and the industrial age under a single anthropological denominator" claims Adorno (1967: 89). Legitimizing such principles by demonstrating their naturalness to the human condition is of course classically bourgeois and ideological. But rarely before or since are we to see archaeology and anthropology playing such a significant role in providing universal and generalizing support.

Extending the entrepreneurial spirit of capitalism into the past now gives a new slant to traditional accounts of barbarian Europe. The stagnant Orient is now held up as the example of what happens when too much surplus is channelled to sustain idle aristocrats or is squandered in dynastic battles and diplomatic gift giving. Prior to the rise of fascism in Europe, prehistorians were prone to debate the positive contribution that the Aryan invasions made to world progress. By contrast the Megalithic builders of Western Europe in the Neolithic displayed all the old orientalist vices. In 1926, the prehistorian Gordon Childe in a book called *The Aryans* could write:

It seems as if these people (the Megalithic builders) were wholly absorbed in the cult of the dead, as if superstitious observances monopolized and paralysed all their activities. Complete stagnation ruled in industry, and to find parallels to their culture we have only to visit the Pacific Islands which may have been exposed to a similar influence (Childe 1926: 211).

The Aryans bring metallurgy and craftsmen who by not being tied to court and aristocratic demand, can produce and trade freely as itinerant workers.

> Thus the Aryans do appear everywhere as promoters of true progress and in Europe their expansion marks the moment when the prehistory of our continent begins to diverge from that of Africa or the Pacific (Childe 1926: 211).

Later Childe was to express revulsion at this Occidentalist thesis but he was not alone in attributing the success of Barbarian Europe to a *longue durée* founded at a particular historical moment. One which invariably involved some combination of Indo-European invasions; large scale replacement of population and language as well as technological change under distinct social conditions. By being recast in vulgar Marxist economic language the thesis concerning the unique development of Barbarian Europe i.e. Europe had been capitalist since the Bronze Age - could continue to dominate European prehistory until quite recently (Rowlands 1987).

Childe in *The Prehistory of Europe* could write

> The metics at Athens, the wayfaring journeyman of the Middle Ages, and the migrant craft unionist of the 19th century are the lineal descendants of the itinerants (bronze smiths) just described. Even in prehistoric times, barbarian societies in Europe behaved in a distinctively European way (op.cit.: 173).

Projected geographically to an ethnographic periphery, civilized identity does not in fact produce a simple primitivist error - a foil to be contrasted with and enjoyed. It is surprising how many of the barbarian thinkers of the early 20th century read widely in ethnography. However, the focus that developed on lavish squanderings of wealth is quite striking, with the Potlatch of the North West Coast Indians figuring large, or the problem of gift giving without apparent return presented as enigma. Veblen is awed by the commodity object as symbol expressing the forceful and superior character of the person possessing it. A primitive mentality is concocted which legitimizes the notion that owner-ship originates in bold acts of hunting, fighting, conquest and plunder. A connection is established between possession and strength, ownership and power, prodigious displays of commodities and a superior position of status.

There is no doubt that this primitive mentality is to be found alive if transformed in the modern mind. Under the common sense barbarian appreciation of worth and honour, the taking of life is the killing of formidable competitors and therefore honourable to the highest degree. "This casts a glamour of worth over every act of slaughter and over all the tools and accessories of the act," claimed Bataille. "Commodities have life because of the capacity of human will to impose itself upon the matter. Things can do for human beings what human beings cannot do for themselves and thus enables them, to become what they are not" (Bataille 1985).

This stress on investment in things is meant to stand in contrast to Marx on the chilling "thinghood' of goods which makes them autonomous and beyond human control. The connection with reification in the writings of the early Lukacs is obvious enough. But the stronger parallel is with the later work of Bataille on sovereignty. To be sovereign means not to let oneself be reduced to the status of an object, as in labour, but to free the subject from bondage. Useless consumption is the best guide on how this is to be achieved based on the principle of whatever pleases me. Bataille bemoans the fate of a past feudal world of brilliance, pomp and wasteful extravagance. In bourgeois society, surplus and wealth have been removed from orgiastic display in the everyday world and become instead sacrificial gifts exhibited on private walls for their prestigious qualities. The progressive expulsion of sovereignty from the realm of useful works has had an evolution of world historical proportions and is still not complete but threatens to be so in a future bureaucratic socialism where power, purified of any associations with sovereignty, emerges unmixed and objective; defined by the imperatives of the system.

The most striking and influential part of all this is Bataille's emphasis on the non-dualism of opposed rationalized and aestheticized worlds. On the one hand the sphere of labour is bounded by limits which banish the violence of an exuberant nature from the ordinary course of affairs. On the other the limits must themselves be prohibitions imbued with fear and horror to resist the temptations of unfettered desire.

That is the nature of the taboo, claims Bataille, it makes possible a world of tranquillity and reason, but it is itself and in its very principle, in the nature of a shudder that befalls not the intelligence but the spirit (Bataille 1979: 59).

Conclusion

At the turn of the century, the development of the sociology of the Durkheimian tradition was to be the critical factor in the transformation of anthropology from evolutionism to functionalism. Its development in the context of a French nation state in crisis is understandable and frequently commented upon. Equally it is often assumed to be a revival of a long tradition that has always emphasized the effect of the collectivity on the individual. But this also has to assume that the collectivity is ordered or rather has a specific form that is compatible with the specific notion of the person that goes with it. Hence, in antiquity, Aristotle could claim that a person is a creature of the city state since to be a non-citizen was to be a non-person. Equally, St. Augustine could define the Christian person as in the image of God. When Durkheim speaks of the person in a putative primitive society therefore it is the conception of the individual in a significant and ordered collectivity. That it should be an idealized 'primitive society' is highly likely, rather in the same way that whilst the spirit of the totally disinterested gift to be found in the ideal tribal society and has only ever been located in state societies with significant commercial sectors. Nor is this altered much in Durkheim's view of modern development. As is well known, his is an uncompromisingly collectivist one. Religion began as nothing but society worshipping itself. Modern societies, he claimed, were moving towards a new kind of sacredness; that of the human individual. This new religion where man is both individual and collective; being and god at the same time neatly disposed of the contemporary *fin-de-siècle* cult of the self by denying it autonomy. How this helps most of us who occupy conflicting roles within collectivities and usually find ourselves deeply at odds with all of them at certain times is something I leave for you to judge but the broader implication is that it is precisely the conflict and potential violence of competing interests that is excluded from concern. Nowhere is this exclusionary tactic more clearly represented than in the anthropology of Marcel Mauss, a pupil of Durkheim and close friend of the surrealist ethnographers (Clifford 1981).

Mike Rowlands

At this point we can return to Lévi-Strauss' adage in the introduction to this paper. He held out the hope that Anthropology
would make us aware that the imperfectly authentic in modern
societies could be expanded in the face of a determined logic to
the contrary. He was optimistic that anthropology could act as an
antidote to what he saw as a tradition of "lawless humanism" in
western culture which treated nature as an object and cultural
diversity as an obstacle to progress. Civilization in the form of a
telegraph pole, a missionary, or an international agency intrudes
everywhere and is a modern barbarism which confronts and
overcomes an ideal primitivism. In his complaints of the increasing separation and alienation of subject and object; state and
society, fact and theory etc. he expressed a long standing anxiety
which took on a particular urgency and pathos at the turn of the
century. Nor has this abated and perhaps we can see a growing
consensus based on Schiller's contention that the domination of
nature must necessarily imply the domination of humans. This
involves a rejection of the humanist Enlightenment privileging
of instrumental reason - both human nature and the natural
world - as something to be harnessed and controlled (Taylor
1986). Yet as we have seen the consensus was not always so one
sided. A pitiless pursuit of a capitalist logic was also feared by
many to be on the wane. What then if in fact anthropology's task
has always been to neutralize and to naturalize the excesses of
capitalist reorganization prior to renewed expansion?

In the period from 1870, Europe's barbarian origins were con-
verted from a source of guilt and shame to honourable status. A
prehistory of an authentic and unique Europe was created which
naturalized an entrepreneurial spirit that had been created in
Europe in the Bronze Age. Whilst this no doubt served many
legitimizing purposes in the major period of European expansion
into a global and colonial phenomenon, it also produced its antithesis eventually in an anthropology that denied individualism
and denied history for a non-western world. The fact that this is
no longer acceptable is no doubt rooted in geo-political realities
that will not subscribe to such one sided readings of cultural
identity any longer. Anthropology has therefore to return to its
European origins to answer its own question of what it is to be
human: the promise held out in its decline as anticipated by
Adorno. The optimistic reworkings of the themes of freedom,
self realization and the refusal of repression and suffering
suggests no lessening of the constraints on their realization. The

short history of barbarism in western culture that I have presented to you would suggest that anthropology's ideological role to expand the authentic in modernity is also an incomplete project.

Bibliography

Adorno, T. (1967) Veblen's attack on Culture, in: *Prisms*, Neville Spearman, London.

Anderson, B. (1983) *Imagined Communities*, Verso, London.

Bataille, G. (1979) *Der heilige Eros*. Suhrkamp, Frankfurt.

Bataille, G. (1985)The notion of expenditure, in:*Visions of Excess* University Press, Manchester.

Beetham, D. (1974) *Max Weber and the theory of modern politics*, Allen Unwin, London.

Benjamin, W. (1966) *Briefe*, eds. Adorno and Scholem, Frankfurt.

Childe, V. G. (1926) *The Aryans*, London.

Dopsch, A. *The Economic and Social Foundations of European Civilisation*, Routledge and Kegan Paul, London.

Engels, F. (1954) A contribution to the early history of the Germans, in: Marx and Engels *Precapitalist socio-economic formations*, Lawrence and Wishart, London.

Kaye, J. (1985) Islamic imperialism and the question of some ideas of "Europe", in: *Europe and its Others*, ed. Francis Barker et al., Colchester, University of Essex.

Levi-Strauss, C. (1963) *Structural Anthropology*, vol. 1.

Le Bon, G. (1896) *Psychologie des Foules* , 2nd ed. , Paris.

Le Bon, G. (1966) (1898) *Psychology of Socialism*, trans B. Miall, London.

Mosse, G. (1966) *The Crisis in German Fascism*, Macmillan, London.

Nye, R. (1973) Two paths to a psychology of social action: Gustav Le Bon and George Sorel, in: *J. Modern Hist.* 45,3: 411-438.

Rowlands, M. (1987) 'Europe in Prehistory": a case of primitive capitalism? in: *Culture and History* vol 1, pp 63-78.

Sorel, G. (1950) *Reflections on Violence*, trans T. E. Holmes and J. Roth, New York.

Spengler, O. (1926) *The Decline of the West*, London.

Taylor, C. (1986) Foucault on freedom and truth, in: David Couzens Hoy (ed.) *Foucault. A Critical Reader* , Blackwell, Oxford.